Who Gets What — and Why

The New Economics of Matchmaking and Market Design

ALVIN E. ROTH

An Eamon Dolan Book
Mariner Books
Houghton Mifflin Harcourt
BOSTON NEW YORK

First Mariner Books edition 2016

www.hmhco.com

Library of Congress Cataloging-in-Publication Data is available.
ISBN 978-0-544-29113-3 (hardback)
ISBN 978-0-544-28839-3 (ebook)
ISBN 978-0-544-70528-9 (pbk.)

Book design by Chrissy Kurpeski
Typeset in Minion Pro

Printed in the United States of America
DOC 10 9 8 7 6
4500794809

PRAISE FOR

Who Gets What – and Why

"If you have a market you want to work better, Al Roth is your man. His new book is fun and compelling – social science at its best."
– N. Gregory Mankiw, Robert M. Beren Professor of Economics, Harvard University, author of *Principles of Economics*

"Roth has now developed a reputation as a troubleshooter in markets as varied as public school placements and residencies for medical school students. This book is his way of sharing what he learned along the way, making it an intriguing field guide from a true pioneer." – *Maclean's*

"Alvin E. Roth is plainly passionate about his subject matter. Thankfully he's also adept at translating the big concepts here into lay language . . . At first glance the ideas in *Who Gets What – and Why* can seem tangled, but tease them apart a bit and you'll find Roth has met his goal. This is heady science that will change your view of the world around you." – *BookPage*

"The co-recipient of the 2012 Nobel Memorial Prize in economic sciences introduces what he calls the new economics of matchmaking and market design . . . Roth's case studies illustrate how problems that obstruct successful matches can be identified economically and overcome . . . An exciting practical approach to economics that enables both individuals and institutions to achieve their goals without running afoul of the profit motive." – *Kirkus Reviews*, starred review

"Practical as well as theoretical. Understanding how matching markets operate can help readers navigate them more effectively. A solid match for readers in general economics and business collections."
– *Library Journal*

To Ben and Aaron, *Emilie,* and Ted

Acknowledgments

Market design is a team sport, so I owe a great debt to all those who have worked on the markets reported here, many of whom are named in the course of telling the story. It turns out that making a book is also more of a team sport than I imagined. I received lots of help on this book, and would have received more if I were easier to help. Worth special mention are my agent, Jim Levine; Tim Gray, who conducted interviews of participants in kidney exchange and school choice; Mike Malone, who made my paragraphs shorter and clearer (and who knows a lot about Sooners); and my publisher, Eamon Dolan, who had clear ideas about what should be included in the book and what should not. I'm also indebted to Barbara Jatkola for thoughtful copyediting, and to Atila Abdulkadiroğlu, Eric Budish, Neil Dorosin, Alexandru Nichifor, and Parag Pathak for careful reading and insightful comments.

Contents

PART I

Markets Are Everywhere

1

Introduction: Every Market Tells a Story

IT WAS 5:00 a.m. on an April morning in 2010. Eight teams of surgeons were preparing to operate on eight patients in four different cities. Four healthy people would each be donating one of their kidneys to someone they had never met, and those four recipients, each suffering from end-stage renal disease, would receive a new lease on life.

At the same time, Jerry and Pamela Green were at their kitchen table in Lincoln, Massachusetts, studying the weather. They were soon to fly as volunteers, in their own small airplane, to Lebanon, New Hampshire, to pick up one of those kidneys, take it to Philadelphia, pick up another kidney there, and take it to Boston. (Two other pilots would transport the other two kidneys.) Because they identified their flight with the call sign "Lifeguard," signifying medical urgency, the air traffic controllers would take them, no questions asked, right through one of the world's busiest airspaces, down the Hudson River and over Newark airport, on their way to Philadelphia, where they would be scheduled to land immediately. Several jetliners carrying hundreds of passengers would be briefly delayed by their passage.

• • •

Kidneys for transplantation are scarce. So is airspace: an airliner uses several hundred dollars per minute in fuel, and only one airplane can occupy a given block of airspace at a time. Passengers' time is also costly. Who got which kidney, which operating room, and which flight path that day in April all required an allocation of scarce resources, so it is perhaps fitting that when Jerry is not flying a small plane, he is a professor of economics at Harvard.

Economics is about the efficient allocation of scarce resources, and about making resources less scarce.

Those kidneys and flights weren't the only scarce resources that had been allocated to bring everything together on that day when four lives were saved. Years earlier, each surgeon had been admitted to medical school and then had proceeded through surgical residencies and fellowships. At each stage, they'd competed with other aspiring physicians. Jerry himself had to go through a similar set of competitions to get his job. Before embarking on their professional training, Jerry and the surgeons had been admitted to colleges, and before that Jerry had been admitted to Stuyvesant, New York City's most selective public high school. Notice that none of these things – kidneys, places in competitive schools, sought-after jobs – can be acquired by the person willing to pay the most or work for the lowest wage. In each case, a match must be made.

Matchmaking

The Talmud tells of a rabbi who is asked what the Creator of the universe has been doing since the creation. The rabbi answers, "He has been making matches." The story goes on to make clear why making matches – in this case, successful marriages – is not only important but also difficult, "as hard as dividing the Red Sea."

Matching is economist-speak for how we get the many things we choose in life that also must choose us. You can't just inform Yale University that you're enrolling or Google that you're showing up for work. You also have to be admitted or hired. Neither can Yale or Google dictate who will come to them, any more than one spouse can simply choose another: each also has to be chosen.

Often there is a structured matchmaking environment—some kind of application and selection process—through which that courtship and choosing takes place. Those matching processes, and how well we navigate them, determine some of the most important turning points in our lives, and many smaller ones, too. Matching dictates not only who gets admitted to the best colleges but also which students get into the most popular courses and which ones live in the best dorms. After college, it determines who lands the best jobs and who has the best opportunities for advancement. Matching sometimes is the gatekeeper of life itself, as when it determines which desperately ill patients receive scarce organs for transplant.

Even if matches are made in heaven, they are found in marketplaces. And markets, like love stories, begin with desires. Marketplaces help shape and satisfy those desires, bringing together buyers and sellers, students and teachers, job seekers and those looking to hire, and even sometimes those looking for love.

Until recently, economists often passed quickly over matching and focused primarily on commodity markets, in which prices alone determine who gets what. In a commodity market, you decide what you want, and if you can afford it, you get it. When buying one hundred shares of AT&T on the New York Stock Exchange, you needn't worry about whether the seller will pick you. You don't have to submit an application or engage in any kind of courtship. Likewise, the seller doesn't have to pitch himself to you. The price does all the work, bringing the two of you together at the price at which supply equals demand. On the NYSE, the price decides who gets what.

But in *matching markets,* prices don't work that way. Going to college can be costly, and not everyone can afford it. But that isn't because colleges raise tuition until only as many students can afford to attend as the college can accommodate—that is, until demand equals supply. On the contrary, selective colleges, high priced as they are, try to keep the tuition low enough so that many students would like to attend, and then they admit a fraction of those who apply. And colleges can't just choose their students; they have to woo them, too, offering tours, fancy facilities, financial aid, and scholarships, since many students are admitted to more than one school. Similarly, many employers don't reduce wages until just enough desperate job hunters remain to fill their ranks. They want the most qualified and committed employees, not the cheapest ones. In the working world, courtship often goes both ways, with employers offering good salaries, perks, and prospects for advancement, and applicants signaling their passion, credentials, and drive. College admissions and labor markets are more than a little like courtship and marriage: each is a two-sided matching market that involves searching and wooing on both sides. A market involves matching whenever price isn't the only determinant of who gets what.

Some matches don't use money at all. Kidney transplants cost a lot, but cash doesn't decide who gets a kidney. In fact, it's illegal to buy or sell kidneys for transplantation. Similarly, airport landing slots involve fees, but that isn't what determines who gets them. Access to public education also isn't priced. Taxpayers support schools precisely so that every child can attend for free. Many people would find it repugnant to allow money to decide who gets a kidney or a seat in a sought-after public kindergarten. When there aren't enough kidneys to go around (and there aren't) or seats in the best public schools (there never are), scarce resources must be allocated by some kind of matching process.

Market Design

Sometimes a matching process, whether formal or ad hoc, evolves over time. But sometimes, especially recently, it is designed. The new economics of *market design* brings science to matchmaking, and to markets generally. That's what this book is about. Along with a handful of colleagues around the world, I've helped create the new discipline of market design. Market design helps solve problems that existing marketplaces haven't been able to solve naturally. Our work gives us new insights into what really makes "free markets" free to work properly.

Most markets and marketplaces operate in the substantial space between Adam Smith's invisible hand and Chairman Mao's five-year plans. Markets differ from central planning because no one but the participants themselves determines who gets what. And marketplaces differ from anything-goes laissez-faire because participants enter the marketplace knowing that it has rules.

Boxing was transformed from brawl to sport when John Douglas, the ninth Marquess of Queensberry, endorsed the rules that bear his name. The rules make the sport safe enough to attract competitors but don't dictate the outcome. In just this way, marketplaces, from big ones like the New York Stock Exchange to little ones like a neighborhood farmers' market, operate according to rules. And those rules, which are tweaked from time to time to make the market work better, are the market's design. *Design* is a noun as well as a verb; even markets whose rules have evolved slowly have a design, although no one may have consciously designed them.

Internet marketplaces have very precise rules, because when a market is on the Web, its rules have to be formalized in software. And now that we can access the Internet from mobile devices, we're never far from a market.

Markets are connected: Internet markets depend on the markets for radio spectrum that have allowed smartphones and other mobile access to flourish where only television and radio used to be.

I've helped design some of the markets and matching processes that I'll introduce in this book. Almost all American doctors, for example, get their first jobs through a clearinghouse called the National Resident Matching Program. In the mid-1990s, I directed the redesign of the NRMP's matchmaking algorithm, which today matches more than 20,000 young doctors with about 4,000 residency programs every year. My colleagues and I helped design matchmaking procedures for doctors later in their careers as well. We also helped design the current system for matching students to high schools in New York City (well after Jerry Green navigated that system) and for schools in Boston and other big cities. The exchanges that Jerry and Pam's flights helped accomplish were arranged by the New England Program for Kidney Exchange (NEPKE), which sprang in part from a design I proposed with two economist colleagues, Utku Ünver and Tayfun Sönmez. In 2004, we helped a group of surgeons and other transplant experts found NEPKE, which used the algorithms we wrote to match donors and recipients, and since then we've helped our surgical colleagues make kidney exchange a standard part of transplantation.

Marketplaces

The first task of a successful marketplace is bringing together many participants who want to transact, so they can seek out the best transactions. Having a lot of participants makes a market *thick*. Making a market thick takes different forms in different markets. To build clearinghouses for kidney exchange, for example, we first had to make the market thick by building databases of patients and donors.

Efforts to keep markets thick often concern the timing of transactions. When should offers be made? How long should they be left open? You can see that even in marketplaces for commodities, from a local farmers' market to a stock exchange. The farmers' market near my old home opens at a fixed time, and if you happen to come a bit early, vendors hesitate to sell you so much as a raspberry beforehand. If they did, they would incur the wrath of their fellow merchants, who worry that if some vendors started to sell before the market officially opened, some customers would come earlier, and an afternoon market could unravel to become an all-day market, requiring the vendors to spend more time selling in a "thinner" market. That's more or less the same reason—to keep the market thick—that the New York Stock Exchange opens for business at the same time each day and closes just as punctually.

Congestion is a problem that marketplaces can face once they've achieved thickness. It's the economic equivalent of a traffic jam, a curse of success. The range of options in a thick market can be overwhelming, and it may take time to evaluate a potential deal, or to consummate it. Marketplaces can help organize potential transactions so that they can be evaluated fast enough that if particular deals fall through, other opportunities will still be available. In commodity markets, price does this well, since a single offer can be made to the entire market ("Anyone can buy a pint of my raspberries for $5.50"), but in matching markets, each transaction may have to be considered separately, as in job markets, in which each candidate has to be evaluated individually.

Although it's great to have a marketplace that gives you an abundance of opportunities, these may be illusory if you can't evaluate them, and they can cause the market to lose much of its usefulness. Think of an Internet dating site on which women with appealing photos receive far more messages than they can answer and men find that very few of their messages draw responses. This causes men to send more, and hence more superficial, messages and women to respond to fewer and fewer of them. Just as women can have more messages than they can answer, employers can have more applicants than they can interview. In both cases, congestion has set in, and that can make it impossible for participants to identify the most promising alternatives the market has to offer.

While buyers like to see many sellers, and sellers like to see multitudes of buyers, sellers aren't so wild about competing with all those other sellers, nor are buyers necessarily glad to have such a crush of competition. So sometimes someone will try hard to transact before the marketplace opens, and in some of the labor markets we'll see in this book, this has led to increasingly early offers or to increased insistence that the offers be answered immediately, before other offers can be entertained. It can be hard to determine when early "exploding" offers are meant to gain an advantage over potential competitors and when they are just attempts to deal with congestion (i.e., if there isn't enough time to make enough offers, start early and move fast). In either case, early exploding offers dilute the thickness of the market and sometimes lead to big reorganizations, such as the development of the labor market clearinghouses for doctors.

One thing that all markets challenge participants to do is to decide what they like. Students have to consider which colleges will suit them, and colleges have to sort through thousands of applications. What often makes matching markets especially challenging is that everyone has to puzzle through not only their own desires but also those of everyone else and how all those other market participants might act to achieve their preferences. College admissions officers aren't simply trying to pick the best students. They're trying to pick the best students who will choose to attend if admitted (and this involves considering where else those students have applied and whom those competing colleges are likely to admit). And so students have to try to signal to colleges not only how *good* they are but also how *interested* they are. Should they apply, via binding early admission, to one school? If so, should they pick the school that they like best but that might be a long shot, or should they apply to a school that's more likely to value their expression of commitment and admit them? In short, both students and colleges have to make decisions that depend a lot on those made by many other students and colleges. (As they say about football, everything is complicated by the presence of the other team!)

Decisions that depend on what others are doing are called *strategic decisions* and are the concern of the branch of economics called *game theory.* Strategic decision making plays a big role in determining who does well or badly in many selection processes. Often when we game theorists study a matching process, we learn how participants "game the system." Well-designed matching processes try to take into account the fact that participants are making strategic decisions. Sometimes the goal of the market designer is to reduce the need to game the system, allowing choosers to concentrate on identifying their true needs and desires. Other times the goal is to ensure that even if some gaming is inevitable, the market can still work freely. A good marketplace makes participation *safe* and *simple.*

When a market doesn't deal effectively with congestion and participants may not be able to find the transactions they want, it might not be safe for them to wait for the marketplace to open if some opportunities are available earlier. Even when going early isn't an option, the marketplace might force participants to engage in risky gambles.

This was the issue that led Boston Public Schools to invite my colleagues and me to help redesign the system for matching children to schools. Under Boston's old system, parents had to strategize about which school they named as their first choice, since the assignment rules made it difficult to get their child assigned to a good school if they didn't list that school first. That wasn't simple. The new system, in contrast, makes it safe for parents to list their true preferences and frees them to think about which schools they actually like best, without having to decide which one school they're prepared to gamble on.

Every market has a story to tell. Stories about market design often begin with failure – failure to provide thickness, to ease congestion, or to make participation safe and simple. In many of the stories in this book, market designers are like firefighters who come to the rescue when a market has failed and try to redesign a marketplace, or design a new one, that will restore order.

But markets can succeed on their own practical terms and still fail in the eyes of those who don't or won't participate in them. Some markets are regarded as repugnant; these run the gamut from slavery to illegal drugs to prostitution. Kidney exchange arose in the shadow of laws around the world that criminalize buying and selling human organs for transplantation (despite which laws, black markets exist, some of which work very badly indeed).

Repugnant transactions—transactions that some people don't want others to engage in—don't have to involve money. Witness the debates on the status of same-sex marriage. But often the addition of money makes an otherwise acceptable transaction seem repugnant, which is why there are laws against selling kidneys but not against kidney exchange, and why consensual sex is generally acceptable but prostitution is generally not. Note, however, that in some places consensual sex (say, between unmarried partners) is considered repugnant. And in some places, prostitution is legal. Repugnance shows with particular clarity what all markets reveal: people's values, desires, and beliefs.

A New Way to See Markets

For me, economics has always had the fascination of gossip: it exposes intimate details of other people's lives and choices. It tells us what kinds of choices we must be prepared to make in our own lives and also which ones we would have faced if we'd chosen a different path.

I hope this book offers insights into matches that you face. Are you trying to get your child into a good kindergarten? Or help her navigate college admissions? Are you applying for a new job? I aim to get you thinking in new ways about navigating those matching processes.

I also hope this book will help you better understand how some forms of organization work well or badly.

I want to shed light on the frequently simplistic assertions we hear from politicians about free markets. Just what is it that allows a market to function freely? When we speak about a free market, we shouldn't be thinking of a free-for-all, but rather a market with well-designed rules that make it work well. A market that can operate freely is like a wheel that can turn freely: it needs an axle and well-oiled bearings. How to provide that axle and keep those bearings well oiled is what market design is about.

Finally, this book—and here is my fondest hope—aims to unveil the economic world in the way that hikes with my friend the Israeli botanist Avi Shmida open my eyes to plants and animals. Once, in the southern Jordanian desert, Avi pointed out a single succulent green plant where the only other growth was dry, dusty scrub. "What do you know when you see a green plant in the desert?" he asked. I shook my head, and he exclaimed: "It's poison! Otherwise something would have eaten it by now."

Another time, Avi commanded me to stick my finger deep into the flower of a sage plant. When I withdrew my finger, it had a line of pollen on the back. Avi then explained how this flower has evolved so that bees have to reach deep inside to get nectar, and thus only big bees with long tongues can get it. The pollen sticks to their backs, where it will be safely transmitted to the next flower they visit. The flower of this plant and bees have coevolved to take advantage of what each offers the other: The flower offers an especially rich source of nectar that can be harvested only by big bees. Big bees, therefore, have a reason to specialize in this kind of flower, which means that the pollen has a good chance of being delivered to another flower of the same species (which is the point of the flower, from the plant's point of view). In this case, evolution has been the matchmaker.

The economic world is just as full of surprising detail as the natural world, and markets also often arise by a kind of evolution, by trial and error, without any intelligent design. But markets can also be designed, sometimes from scratch but often after trial and error leads to a market failure. Much of what we've learned about market design—and from market design about markets more generally—has come from observing market failures and figuring out how to fix them. Not all markets grow like weeds; some, like hothouse orchids, need to be nurtured. And some carefully nurtured marketplaces on the Internet are now among the world's biggest and fastest-growing businesses.

Like flowers of different species, marketplaces for different kinds of goods and services are often quite different from one another. But, also like different species of flowers, even very different marketplaces have some things in common, since they arose from a need to solve similar problems.

When I look into markets that are suffering from some sort of failure, not only do I get to see how other people's lives unfold at some of their most important junctures, but I also get to meet an exciting cast of characters whom I'd like to introduce you to. Because economics touches on just about everything, economists have an opportunity to learn something from just about everyone, and I've met and worked with some remarkable people in each of the markets I've helped design.

Market design is giving new scope to the ancient profession of matchmaking. Consider this book a tour of the matching and market making happening around us. I hope it will give you a new way to see the world and to understand who gets what—and why.

2

Markets for Breakfast and Through the Day

MARKET DESIGN IS so pervasive that it touches almost every facet of our lives, from the moment we wake up. The blanket you chose to sleep under, the commercial playing on your clock radio —even the radio itself—embody the hidden workings of various markets. Even if you eat only a light breakfast, you likely benefit from the global reach of multiple markets. And while most of those markets are easy to participate in, even that apparent simplicity may disguise a sophisticated market design.

For example, you probably don't know where your bread was baked—but even if you do, your baker doesn't have to know who grew the wheat that went into the flour used to make the bread. That's because wheat is traded as a *commodity*—that is, it is bought and sold in batches that can all basically be considered the same. That simplifies things, although even commodities need to be designed, so that the market for wheat doesn't have to be a matching market, as it was as recently as the 1800s.

Every field of wheat can be a little different. For that reason, wheat used to be sold "by sample"—that is, buyers would take a sample of the wheat and evaluate it before making an offer to buy. It was a cumbersome process, and it often involved buyers and sellers who had successfully transacted in the past maintaining a relationship with one another. Price alone didn't clear the market, and participants cared whom they were dealing with; it was at least in part a matching market.

Enter the Chicago Board of Trade, founded in 1848 and sitting at the terminus of all those boxcars full of grain arriving in Chicago from the farms of the Great Plains.

The Chicago Board of Trade made wheat into a commodity by classifying it on the basis of its quality (number 1 being the best) and type (winter or spring, hard or soft, red or white). This meant that the railroads could mix wheat of the same grade and type instead of keeping each farmer's crop segregated during shipping. It also meant that over time, buyers would learn to rely on the grading system and buy their wheat without having to inspect it first and to know whom they were buying it from.

So where once there was a matching market in which each buyer had to know the farmer and sample his crop, today there are commodity markets in wheat, corn, soybeans, pork bellies, and numerous other food items that are as anonymous—and efficient—as financial markets. Just as investors don't worry about which particular shares of AT&T stock they buy, buyers don't care which particular 5,000 bushels of number 2 hard red winter wheat they have shipped to them. Thanks to the rating system, they can buy wheat without seeing it. Commodifying wheat via a reliable grading system helped make the market safe.

Wheat can even be sold *before* it's harvested, as *wheat futures*—a promise of wheat to come. This allows big millers and bakers to make their purchases and lock in their costs in advance. They can do so without fear, because the standardized description of what is being purchased means they don't have to worry about what will be delivered. The purchase of wheat futures is a purely financial transaction, with no wheat even present in the marketplace.

As for the transaction itself, brokers inspecting and buying lot by lot have been replaced by commodity traders on the floor of the Chicago Board of Trade signaling and calling out their bids and offers in the trading pits of the open outcry markets that came to dominate this kind of transaction. Nowadays traders also buy and sell enormous volumes of grain while sitting at computer screens.

Turning a market into a commodity market helps make it really thick, because any buyer can buy from any seller, and any seller can sell to any buyer. At the same time, it also helps the market deal with one of the main sources of congestion in matching markets, since in a commodity market each offer to sell can be made to all buyers, and each offer to buy can be made to all sellers. So unlike in the market for jobs, or for houses, no one has to wait for an offer to be made to him personally; anyone who sees (or hears) a price he likes can take it. We'll see in more detail how such markets can work when we look into financial markets in chapter 5, and we'll see just how fast commodity markets can sometimes operate.

Coffee and More

Turning a product into a commodity can affect not just how it's bought and sold but even what is produced. Still keeping our sleepy eyes squarely on the breakfast table, let's shift our attention to coffee and its own remarkable market tale.

Coffee beans have been grown in Ethiopia for centuries, but until the twenty-first century they were traded a lot like nineteenth-century American wheat. If you wanted to buy Ethiopian coffee in bulk at the source, you had to have an agent there who could extract a sample from deep inside each sack to taste and evaluate it.

That changed in 2008 with the creation of the Ethiopia Commodity Exchange. At its heart is a system of anonymous coffee grading, in which professional tasters sample and grade each lot put up for sale. (By the way, there was also some thoughtful market design that went into the rules — that is, the market design — involved in organizing quality grading. For example, tasting must be "blind"; the tasters can't know whose beans they're tasting. Otherwise they could be bribed by the seller to inflate the grades.)

The standardization of coffee can actually improve the quality of the coffee harvest. Coffee beans grow inside a "cherry," and the best coffee is harvested when the cherry is ripe and red. But the beans are sold after being removed from the cherry and dried. So when buyers simply see coffee beans, they can't tell whether they were harvested from ripe red cherries or from unripe green ones. Before coffee was graded, coffee farmers sometimes were tempted to harvest a whole hillside at once, red and green beans, ripe and unripe. But now that tasters can tell the difference, it makes sense to have coffee pickers pluck only the red cherries and to come back later to harvest the rest of them when they are ripe. Since the graders can tell the difference, the market reliably rewards such care with a higher grade and a higher price. The ultimate result is that foreign buyers can now buy Ethiopian coffee beans in bulk from a distance, without having to taste them on the spot, and from multiple sellers, without worrying about the sellers' reputation or pedigree.

So as you sip your morning coffee, you are benefiting from some fairly recent design in the marketplace for an ancient agricultural commodity, which wasn't always as standardized—or as good—as it is today.

That said, your coffee doesn't necessarily come to you anonymously, even if you don't know who grew the beans. You may run out to pick up your coffee already brewed from Starbucks or a more local coffee shop, but in either case you know quite a bit about the seller. You may have chosen your coffee joint for its convenience, for the pastries it sells with the coffee, or even for the designs the barista swirls into the foam on your latte. And if you're a regular, that seller may also know a lot about you—for instance, getting your "usual" ready when she sees you walking in.

Coffeehouses try hard to differentiate their products so that customers will want to return and buy regularly from them. Of course, if you're in a strange city, you may find yourself seeking a big chain such as Starbucks precisely because of the standardization of the drinks it sells, since you haven't had a chance to locate a more idiosyncratic coffee shop that might suit you better.

Notice the tension between commoditization and product differentiation—that is, between wanting to sell in a thick market to buyers even if they don't care who you are, and trying to make your product special enough that many buyers will care enough about you to seek you out. Sellers enjoy selling in a thick market of buyers, but they don't enjoy being interchangeable with other sellers. Giant brand leaders such as Apple and Microsoft sell products that are enough like commodities that you don't care which particular iPhone or copy of Microsoft Office you have, but they are different enough that you can't buy the same phones and software from anyone else. Part of Apple's success is that it sells a unique brand of laptop computers, while the PCs pioneered by IBM became a commodity that could be sold by other companies as well. This opened the door to Microsoft's near monopoly on the operating system that runs all PCs, since their spread created a big, thick market for software on the PC platform.

In much the same way, there's a tension between commodity markets and matching markets. You care who brews your coffee, but your coffee shop sells to all comers. That is, in the market for a cup of coffee, your coffee shop has to be chosen, but you get to choose—and you care whom you choose. So the distinction between perfectly anonymous commodity markets and relationship-specific matching markets isn't a thin bright line. Rather, there are markets at different points along a spectrum from pure commodity to pure matching. When I buy bread in the supermarket, I don't really know the baker, but I can recognize that it's the usual bakery, since the baguettes I get come with the bakery's name printed on the bag, along with the information that it has been cheerfully baking bread since 1984.

Buyers have some of the same ambivalence as sellers: while we like the fact that some goods are commodities that we can buy without inspecting, we also enjoy variety and seek out unusually high and hard-to-standardize quality. Sometimes on Sunday mornings, my wife and I buy our breakfast at a local farmers' market—an ancient format that still attracts busy city dwellers. It's an attractive place to shop, not least because of the perceived freshness available in a marketplace that is open only one day a week. You know for sure that the goods came to the market that day and didn't languish in a supermarket's storage room before being put on the shelf.

Moreover, the farmers showing their wares are typically local. And because the farmers themselves (or their families) are usually manning the stand, you can easily find out something about them. The result is more of a matching market than when you stock up in your local grocery store, although that store is open every day, which makes it more convenient.

The grocery store may be open every day, but it's not open all the time, because it's costly to keep a store open when there are only a few potential shoppers. But whether you shop at the farmers' market or the local supermarket, you still have to go there to make your purchases. The Internet is changing all that and making markets more ubiquitous.

Marketplaces in the Air . . . and Everywhere

These days, with your smartphone and your credit card, you can buy a plane ticket, make a hotel reservation, order a meal to be delivered, or purchase a pair of shoes. On the Web, you can buy from millions of different sellers—and if you point the browser on your phone or your computer at a big Internet marketplace such as Amazon, you can fill your virtual shopping cart with items from multiple sellers and buy them in a single transaction. That's part of what makes Internet markets so easy to use and so successful. When my watch breaks, I might go to Amazon to buy a new one. But I might also buy a mirror for my bike helmet and a book I've been planning to read, then pay for them all with a credit card and have them shipped to my home. It looks to me like a single transaction, even though I may have bought each item from a different seller that subscribes to Amazon's marketplace services.

In attracting so many shoppers and so many merchants, Amazon has created a thick marketplace, one in which there are many participants ready to make many different kinds of transactions. The thickness of the Amazon marketplace—the ready availability of so many buyers and sellers—is self-reinforcing. More sellers will be attracted by all those potential buyers, and more buyers will come to this marketplace because of the ever-expanding variety of sellers. So Amazon lets me shop easily for many different things in the same place, and my phone lets that place be wherever I am.

Your smartphone is a marketplace not only for goodies from Amazon but also for software applications, or apps, that expand what your phone can do. That's why your phone almost certainly runs on one of the two most popular smartphone operating systems, Apple's iPhone or Google's Android. People want phones with a long list of apps to choose from, and they know that they'll want some apps later that haven't even have been invented yet. At the same time, a software developer writing an app wants to sell it in a marketplace with lots of potential buyers so the app will have a chance to become a big hit.

Phone buyers and app developers are looking to meet in a thick marketplace—one with many possibilities on the *other* side of the market. That's why independent developers first write apps for phones with many users, and phone buyers look for phones with an abundance of apps. Your phone's operating system is the key to the marketplace, since each app has to be written to be compatible with a particular operating system.

Apple and Google both launched their proprietary operating systems with a multitude of apps already available so that customers would be attracted immediately by their thickness. But Apple and

Google made other, notably different choices when designing their markets. Apple chose a "closed" operating system that allowed it to control which apps could be sold to iPhone users. Google, which came later to the game, opted for an "open" system, publishing the code so that any developer could build for it. These choices echoed similarly opposing strategic decisions made by Apple and Microsoft at the dawn of the personal computer age. Anybody could make software for the PC platform, but only Apple (or those developers it allowed to do so) could make software for its personal computer, the Mac. These choices allowed the market for PC software to grow thick much more quickly than the market for Mac software. But Apple's decision to keep both its hardware and software on a proprietary standard eventually allowed it to reap huge profits.

As with other kinds of markets, popular operating systems quickly get more and more popular, as they attract both new buyers and new sellers. In time, they become de facto *industry standards*—meaning they essentially establish a marketplace in which products (new applications) can be sold. Once this happens, they can, at least for a time, so completely dominate their markets that competing operating systems can't attract enough users and developers to be anything but niche offerings.

That's exactly what happened in the smartphone market. The two most popular operating systems, iPhone and Android, have captured so much of the market that they've become almost self-perpetuating. In the process, they have displaced earlier popular Internet phone operating systems, notably the BlackBerry, which in turn had replaced non-Internet phones and non-phone digital assistants such as the PalmPilot.

Notice how markets interact with one another. Amazon couldn't have become the marketplace it is without the Internet, which couldn't have become a marketplace without first computers and then smartphones. And smartphones couldn't have become marketplaces without a way to pay for purchases over the phone. At the farmers' market and the supermarket, anyone can pay in cash if they want to. On the Internet, it's convenient to pay with a credit card. And a credit card is also a marketplace, which is why there's a good chance you have one of the big ones: Visa, MasterCard, or American Express. Consumers who use credit cards and merchants that accept them are all looking for a thick market, with lots of participants on the other side.

I'm old enough to remember when people paid for most things by cash or check. It was hard to pay by check if you were away from home, since merchants didn't like to take the risk that your check would bounce and they wouldn't get paid. But if you were a regular at a local restaurant, the owner was usually glad to take your check — although even then you'd sometimes see a sign over the cash register that read IN GOD WE TRUST; ALL OTHERS PAY CASH.

Credit cards offered merchants safety, but that safety came at the cost of transaction fees. Most merchants were willing to pay those fees because accepting credit cards brought in customers they might otherwise have missed, and also because credit cards made it safe for them to take noncash payment from customers they didn't know well, since the bank guaranteed payment as a form of insurance.

It took a while for the markets facilitated by credit cards to become thick by settling on just a few major cards, but it is hardly surprising that this happened. Imagine how much less useful credit cards would be if the markets had moved in the other direction and every store used a different one. In the early days, some people carried several credit or charge cards, and various businesses accepted only certain ones. This sometimes led to embarrassing moments when the check was delivered at a restaurant. So the cards that were most popular became the most useful ones to carry and to accept, since they gave access to the thickest markets — that is, to the most restaurants and shops on one side, and the most diners and buyers of other goods and services on the other. By the late 1960s, an industry shakeout had already begun. A number of famous cards — most notably Diners Club, which was the first credit card in widespread use — faded into the background.

Part of what makes credit cards work is that they simplify transactions for both buyers and sellers. Concentrating on just a few cards further simplifies matters on both sides of the market. Thus ever since the big shakeout, no new credit cards have joined the ranks of the majors; the barrier to market entry has proved to be too great. That said, in recent years the Internet revolution has opened the door to competition from wholly new directions – including new kinds of payment services, such as PayPal; an international network of automatic teller machines to challenge old standbys such as traveler's checks; and maybe even new types of "virtual money" such as Bitcoin. As I write this in 2014, Apple has announced a new payment system on the latest iPhones, and we can reasonably expect that it and/or other new payment systems that make use of mobile devices will become commonplace.

The bank that handles Amazon's transactions, or the one that manages the account of your favorite restaurant, is typically different from the bank that issued your credit card and takes your payment. So behind the scenes, there is an interbank market, too, through which payments flow. This hidden market eases the congestion that could otherwise result from settling very large numbers of relatively small transactions, in the same way that Amazon itself eases the congestion of making several little purchases from different sellers. This interbank market lets each merchant deal with just one bank, just as your monthly credit card statement enables you to make a single payment that settles your account with many merchants.

Your credit card also acts as a lender. (That's what distinguishes credit cards from charge cards, which offer only the ease of a cashless transaction.) It offers you access to the market for credit, so any time you want to buy something, you can borrow money, though typically at an exorbitantly high interest rate, simply by not paying the full amount you owe when your bill arrives. The bank that issued your credit card can get away with such high rates because once you've made your purchase, the bank isn't facing a lot of competition in offering you easy credit. In fact, you might have chosen this card because it provided cash back on some purchases. It turns out that lots of people who do that never pay much attention to the interest rate, because they're planning to pay their bills in full. But then they seldom switch cards. So there isn't much pressure on banks to lower their rates. I hope you don't borrow on your credit card very often: it's a bad deal—the kind of deal you're likely to be offered when the other side of the market isn't thick.

In thicker markets, where customers have ready alternatives, it's harder for a seller to get away with such bad deals. At one time, merchants tried to pass on the cost of credit card purchases to consumers by charging a premium for using the card instead of cash. This didn't catch on, in part because credit card purchasers disliked it so much and could take their business elsewhere. Instances in which consumers recoil from offers that strike them as unfair are more common than you might think. Even marketing giants are sometimes surprised by what they *can't* get away with. In 1999, for example, Coca-Cola tested vending machines that could automatically raise prices in hot weather. The backlash was quick—and the company abandoned the idea just as quickly. So regular folks who find certain transactions particularly distasteful do have some recourse when they can take their business elsewhere or simply withhold it—and this, too, plays a role in shaping markets.

Incidentally, the fact that most purchases cost the same whether they are paid for by credit card or by cash opens the door to an attractive-looking kind of competition among credit cards that may not be as attractive as it seems. Many credit cards now compete on how much "cash back" they offer to consumers. Those refunds come out of the fees that credit card companies charge to merchants and are reflected in the prices that merchants charge their customers. So when two customers stand in line at the cash register with identical purchases, and one pays with a credit card and one pays cash, the one who is paying cash is paying for the discount that the credit card customer is receiving. That is, as more consumers are attracted to higher cash-back deals, and as credit cards successfully compete for customers by raising these kickbacks, merchants pay larger credit card fees and raise prices in response. And a discount from a higher price isn't such a good discount, especially for those who are paying cash. To put it another way, we pay a cost for the convenience of using a middleman, and that is partly because the middlemen — in this case, the credit card companies — compete for our business in a way that mutes the price competition among merchants that might otherwise bring prices down. It's something to remember: competition can take many forms, and it isn't always easy to see who gains and who loses.

Each of these ubiquitous marketplaces has found a way to succeed not only in making markets thick, uncongested, and safe, but also in making them *simple to use.* Making a market simple to use, however, may not be *simple.* Behind Amazon's one-stop shopping, for example, are storage and shipping, fast Web servers, and secure ways of paying, with encrypted credit card numbers on file so that regular customers don't have to be troubled each time they make a purchase.

Simplicity is a competitive tool that sometimes allows new market platforms to displace old ones. Credit cards replaced paper checks, and it remains to be seen whether mobile payment systems will replace credit cards. If they do, it will be because it's simpler to swipe your phone than your credit card, more secure, or simpler for the merchant to accept payment that way. Notice that when competition among marketplaces causes previously successful markets to fail, it is often the result of undermining the previous success in establishing a thick market. If, for example, mobile payments turn out to be more attractive to merchants than credit cards, then as the mobile payment market becomes thick, some merchants might stop accepting credit cards that charge them a high fee. That would in turn make those credit cards less attractive to consumers, which would make them unattractive to even more merchants, and a previously thick market would start to become thin.

In the chapters to come, you will begin to see markets in sharper focus, with more attention to the details of how they work, the "rules of the game."

A few of the marketplaces I'll tell you about are ones that I've helped design or that I've studied carefully. Others are just markets that I participate in, as you do – such as the market for phones, credit cards, or that morning cup of coffee.

When we think of markets, most of us typically imagine the stock exchange, or a retail shop offering products to customers, or the surging demand for new smartphones, or maybe just a traditional farmers' market. But as we've already seen, we encounter many other markets every day, and our world would be utterly different (and a lot less pleasant) without them. These markets include not only our experiences at the supermarket or phone store but also those in getting into college, finding a job, eating breakfast – even getting a kidney transplant.

One thing we'll see is that the "magic" of the market doesn't happen by magic: many marketplaces fail to work well because of poor design. They may fail to make the market thick or safe, or to deal with congestion, and so there's an opportunity to help them work better. And sometimes there's an opportunity to build a marketplace from scratch, to serve an entirely new market, to facilitate a new kind of exchange. We'll see that in the next chapter, where I tell you about kidney exchange.

3

Lifesaving Exchanges

DR. MICHAEL REES was tired of watching his patients suffer and die.

Too often, that's what happened when he told someone with kidney failure that he would have to wait until a cadaver kidney became available. What made that conversation even harder was that so many patients had come to him filled with hope. They'd already found someone—a family member, a close friend, sometimes just an acquaintance—who was willing to donate a kidney to them (a donor, like any healthy person, needs only one of the two kidneys she was born with). A timely donation can not only save people with kidney disease from the long wait for a deceased-donor organ but also spare them the grueling downward spiral of dialysis.

But a willing donor isn't enough. Blood types have to be compatible, and a patient's immune system must not immediately reject the new kidney. Time after time, Mike did those tests at the University of Toledo Medical Center, only to give his patients the bad news that none of their prospective donors was compatible. He hated that conversation. He'd become a doctor to cure people, not to make them stand in line at death's door, waiting for some other unfortunate person, with sound kidneys, to die.

Then, in early 2000, Mike heard that a kidney "exchange" had been conducted at Rhode Island Hospital. The transplant team, led by Anthony Monaco and Paul Morrissey, had found itself with two incompatible patient-donor pairs and noticed that each donor's kidney would work for the other patient. With the patients' and donors' permission, they did the swap.

Wondering if he might help his patients with similar exchanges, Mike carried home two boxes of patient and donor charts. After putting his kids to bed, he sat at his kitchen table and spent the next four hours poring over them, noting each patient's blood and tissue incompatibilities. Soon charts covered the table. One by one, he compared each patient chart with all of the donor charts. "I didn't really have a strategy," he recalls. "I stayed up that night until I figured out two pairs that might match."

Because of advances in immunosuppression drugs, which reduce the chance that a person will reject a donated organ, a person can receive a kidney from someone who isn't an identical twin or even a blood relative. But finding a match is harder than just getting the right blood type. The fact that my wife and I are parents, for example, reduces the likelihood that she could accept one of my kidneys. During childbirth, she might have been exposed to some of my proteins that our children inherited, and her immune system might have developed antibodies against them.

That was what happened with one of Mike's potential exchanges. Although the exchange looked like it would work because the blood types of the patients and donors were compatible, one of the patients had antibodies against some of the proteins in the proposed donor's kidney. That transplant wouldn't work, and hence that exchange couldn't be done.

Mike's first attempt had failed, but he realized that a kidney exchange could work. What he needed was a big enough database of patient-donor pairs to improve the odds, as well as software that could evaluate the potential combinations. With both of those factors in place, Mike was certain he'd find matches.

Kidneys and cadavers may seem out of place in a discussion about markets. But the story of the creation of kidney exchanges—in which I played a major role—touches on almost every subject I will discuss in the chapters to come, about how market design has to solve problems related to incentives, thickness, congestion, and timing, and how some kinds of transactions can be widely seen as repugnant. In describing how the marketplace for kidney exchange was created, I will be introducing the major themes of this book.

Just as important, the very fact that something as intimate, personal, and, frankly, disturbing as the exchange of human kidneys can not only be organized as a marketplace but can be made better, fairer, and more efficient in the process underscores the first thing I hope you'll start to notice all around you. It is that *markets and marketplaces come in many forms, some of which don't conform to conventional notions of markets, and some in which money may play little or no role.*

So let's return to our story of Dr. Michael Rees's hopes for kidney exchange and use it as an introduction to the design of markets and marketplaces.

When we see a long line of people waiting to buy some scarce good, we suspect that demand exceeds supply. If we know a little bit about economics, we also may conclude that this imbalance is occurring because the price is too low to generate more supply.

As I write this, more than 100,000 people are waiting for a kidney transplant in the United States. Meanwhile, the price of kidneys is zero, since it's illegal here and in most of the rest of the world to buy or sell kidneys for transplant. Sure, lots of money must be spent for hospitals, doctors, and drugs before a transplant can happen. But by law, the kidney itself must be a gift.

So kidneys must be exchanged without money changing hands, in a kind of barter transaction.

In the late 1800s, the economist William Stanley Jevons pointed out that the invention of money was a market design solution that overcame a major problem that severely limited barter, namely the need to find someone who both has what you want and wants what you have. Money eases the need to find this "double coincidence": with money in the market, it's enough to find someone who has what you want. You can buy what you want from that person without having to find someone with whom you can trade goods.

The difficulty that Mike Rees found when he tried to arrange his first exchange was precisely the one Jevons pointed to: no exchange could happen without a double coincidence. The question then became, how do you design a clearinghouse for kidney exchange that can function as an efficient marketplace, but without using money?

Trading Cycles

I was a newly minted game theorist when I arrived at the University of Illinois in 1974. I had just graduated from a Ph.D. program in operations research at Stanford University. Early in my studies, I learned that most of the mathematical tools available for organizing operations focused on things, not people. The kinds of mathematical optimization developed for organizing factories and warehouses and for scheduling freight trains and passenger planes didn't address the fact that different people may have different goals that might have to be accommodated. The exception was the just-emerging field of game theory—the study of strategic interactions. I gravitated toward game theory because I cared about how people made choices and organized themselves. Game theorists try to put themselves in the shoes of market participants to understand how they might use the strategies that are available to them.

That same year, two veteran game theorists, Lloyd Shapley and

Herb Scarf, published an article in the very first issue of the *Journal of Mathematical Economics* in which they posed a thought experiment: *How can people trade indivisible goods if everyone needs just one, has one to trade, and can't use money?* Though Shapley and Scarf didn't have any particular market in mind, they called the goods "houses." As will become clear to you—as it eventually became clear to me—the people in their thought experiment could be incompatible patient-donor pairs, with each pair needing a kidney and having a kidney to trade.

But I was far from thinking about kidney exchange in 1974. Although thought experiments like this one can in time turn into practical tools, they start life as toys. Just as children prepare to be grown-ups by playing, an abstract mathematical model allows economists to play with possibilities in a simplified, uncomplicated way. So Shapley and Scarf had proposed a new toy that could be used to explore how exchange might work in a hard case where you couldn't use money and trade had to be one for one, because everyone had one indivisible item to trade—that is, you couldn't trade something for just a part of something else.

Such trades can take place in *cycles*. The simplest kind of trade would be a two-way cycle, between two patient-donor pairs in which each donor was compatible with the patient in the other pair. A bigger cycle, among three pairs, would accomplish an additional transplant, with the donor from the first pair giving a kidney to the patient in the second pair, the donor in the second pair giving a kidney to the third pair, and the donor in the third pair giving a kidney to the first pair, thus closing the cycle.

Shapley and Scarf showed that for any preferences that patients and their surgeons might have regarding which kidneys they would like, there was always a way to find a set of cyclical trades they called "top trading cycles," with the property that no group of patients and donors could go off on their own and find a cycle of trades that they liked better. Organizing trades this way would help make it safe for surgeons to enroll their patients in such a market, since the patients couldn't do better by trading differently among themselves.

As I began to play with this model, I started to think of it as the potential architecture for a centralized clearinghouse that could help traders overcome the obstacles to barter. But for such a clearinghouse to find the most desirable set of trades, it would need to have access to patients' needs and preferences, and so participation would have to be safe in another way, too.

Since preferences are by and large private information, for a clearinghouse to work people would have to reveal this information. But patients and their doctors might worry that if they told the clearinghouse too much, the clearinghouse might use that added information to give them a less desirable kidney because they were willing to accept it, even when one they preferred was also available. Or they might worry that by trying and failing to get their most preferred outcome, they would lose their chance to get a kidney that was almost as good, because it wasn't their first choice. In 1982, however, I was able to show that top trading cycles made it possible to organize a clearinghouse in such a way as to guarantee to patients and their surgeons that it was safe for them to be completely candid in revealing this kind of information.

Also in 1982, I moved to the University of Pittsburgh, which had the most active organ transplant center in the country. Its director, Thomas Starzl, who'd performed the first successful liver transplant, was a local hero. I used to see him, surrounded by younger surgeons, at the coffee shop near campus. That put organ transplants at the front of my mind. When I taught about trade in indivisible goods without money, I started using kidneys as an example of what was being exchanged instead of Shapley and Scarf's "houses."

Kidneys were a better example than houses because, in the real world, houses are actually traded for money, but it's against the law to use money to trade kidneys. While students are prepared to tolerate being taught about "toy" models, they're happier when they can see that these simple models might have a practical application. And although I am a great believer in the value of abstract models,

I'm also happier when I can see where my work might possibly be heading.

In 1998, I moved to Harvard. Shortly after that, in 2000, the first kidney exchange in the United States took place. In the meantime, progress on another problem had laid the groundwork for my further thinking about kidney exchange. Two Turkish economists, Atila Abdulkadiroğlu and Tayfun Sönmez, had been looking at the problem of dormitory room allocation — yet another problem where money doesn't play a central role.

Allocating college dorm rooms has more in common with organ exchange than you might think. Some students — freshmen — don't have a room and need one. On the other side, there are rooms that have been vacated by graduating seniors that don't have an occupant. There are also rooms that have an occupant who is interested in trading up for another room he prefers. Now apply this to kidneys: Patients with incompatible donors are like occupants who'd like to trade. Patients without a living donor are like roomless freshmen. And deceased-donor kidneys are like the rooms vacated by seniors.

In 2002, a former Ph.D. student of mine from Pitt, Utku Ünver, came to Harvard from Koç University in Istanbul as a research fellow. I suggested we give a lecture on kidney exchange for my market design course. We posted our notes on the Web, and Tayfun, Utku's colleague at Koç, read them and suggested that he join us to collaborate on designing practical kidney exchange.

Our collaboration was intense and tiring, but also exhilarating. The seven-hour time difference between Istanbul and Boston made it seem as if we were working around the clock. When we finished, we had designed an algorithm both for kidney exchange among patient-donor pairs and for integrating these exchanges with "non-directed donors," such as deceased donors (and a growing number of living donors) who'd volunteered to give a kidney to someone in need but who weren't paired with a particular intended recipient.

An exchange that begins with a non-directed donor is a *chain* rather than a cycle, since it doesn't have to return to its beginning: the non-directed donor is an altruistic person who arrives without a patient and is prepared to give a kidney without receiving one in return. In the past, deceased and other non-directed donations had always been directed to someone at the top of the waiting list for deceased-donor kidneys. But kidney exchange now made it possible for a non-directed donation to spark more transplants, since a chain could start with the non-directed donor, include some patient-donor pairs, and end with a donation to someone on the waiting list.

Our algorithm found both top trading cycles among patient-donor pairs and chains that began with non-directed donors, in a way that made it safe for patients and their surgeons to participate. Now all we had to do was turn theory into practice and convince surgeons that we could help them. That wasn't so easy. Doctors don't automatically think of economists as fellow members of the helping professions.

We posted our paper on the Web and sent copies to kidney surgeons across the country. At first only one doctor, Frank Delmonico, a Harvard surgeon and the medical director of the New England Organ Bank, responded. Frank and I began a series of conversations about the logistics of organizing exchanges among many patient-donor pairs.

Frank began by pointing out that the large cycles and chains we proposed would be too complex; he didn't think kidney exchange would be practical among more than two pairs at a time, at least not yet. Because the transplants and nephrectomies (kidney removals) needed to be done simultaneously (more on this later), even an exchange with just two pairs would require four operating rooms and surgical teams. Frank worried that anything larger would be too complicated logistically.

So we went back to work and developed another algorithm. We again paid close attention to the fact that patients and donors can't be scheduled like freight trains and just told where to go. This algorithm, too, had to make it safe for patients and their surgeons to share all necessary information. A lot of this information, however, isn't available automatically; it has to be volunteered.

For example, a well-designed algorithm would elicit information about how many willing donors a patient has. Suppose a patient had two potential donors, his wife and his brother. The algorithm would likely find more potential matches if the patient enrolled in the program with both of those potential donors, even though only one would need to donate. Two donors would increase the chance of being matched with another patient-donor pair, since either of these donors might be the one compatible with the other patient. (By comparison, a matching algorithm that gave priority to patients with only one donor would be bad market design, since patients who had two donors might then reveal only one, so that they could receive priority, too.) So a kidney exchange algorithm had to make it safe for patients and their doctors to reveal all sorts of information, as information is essential to finding the best set of exchanges.

With Frank's backing, this new proposal got a wider hearing. Out of that work came, in 2004, the New England Program for Kidney Exchange (NEPKE), which began by organizing the fourteen kidney transplant centers in New England to help incompatible patient-donor pairs find matches. A year would pass before NEPKE could start doing exchanges. Consent forms had to be obtained from patients and their donors so that databases could be assembled. NEPKE also had to hire staff, including its clinical program manager, Ruthanne Leishman, a nurse with a master's degree in public health. Ruthanne would painstakingly coordinate all the details of kidney exchanges —a complicated task.

Tayfun, Utku, and I were gratified to see our software being used to good effect at NEPKE and also to see the underlying ideas being adopted elsewhere. But we were frustrated to see only two-way exchanges happening. We knew more patients would get transplants if hospitals tried larger exchanges. That's because some patients who might not fit into any two-way exchange might fit into one that involved more pairs. We knew that this was logistically possible because a small handful of successful three-way and even four-way exchanges had already taken place.

In 2005, we wrote a paper showing that many of the advantages of larger exchanges could be captured if transplant centers could regularly conduct exchanges among three patient-donor pairs as well as between just two pairs. (How many more transplants could be done with combinations of two- and three-way exchanges, instead of just two-way exchanges, would depend on the patient pool and how many easy-to-match pairs it contained.) Once again, the time zone difference worked to our advantage, because Utku had now returned to Turkey, while Tayfun had come to Harvard as a research fellow. We proposed a way to organize exchanges that put a limit of three or four pairs that could be included in an exchange. We distributed the paper widely, and this time our colleagues at NEPKE were convinced. Within a year, NEPKE and other networks had incorporated larger exchanges into their operating procedures.

This sounds rather abstract, but recounting one actual three-way exchange will give a clearer — and more dramatic — sense of the process and its impact. In this exchange, the three patient-donor pairs were all married couples living in New England. One donor also happened to be a nephrologist — a physician who treats kidney disease. That donor, Dr. Andy Levey, of Tufts Medical Center, is married to Dr. Roberta Falke, who is an oncologist.

Several members of Falke's family had the same ailment she did: polycystic kidney disease, or PKD. The disease had killed her father at age fifty-four. Two of her four siblings had it, as did Falke and Levey's adult son. Another brother had already donated one of his kidneys to one of Falke's sisters. Several friends volunteered to donate, but none was compatible. Although Levey was healthy enough to donate, his kidney wouldn't work for his wife.

Peter and Susan Scheibe of Merrimack, New Hampshire, and Hai Nguyen and Vy Yeng of Revere, Massachusetts, had endured similarly futile searches. Diabetes, a common culprit in renal failure, was destroying Peter Scheibe's and Hai Nguyen's kidneys. Their wives were willing to donate but were incompatible. NEPKE matched the three pairs, showing that Levey's kidney would work for Peter, Susan Scheibe's kidney would work for Nguyen, and Yeng's kidney would work for Falke. The donations and transplants took place on December 15, 2009.

Levey and Falke had their surgeries at Tufts. Levey found the experience of being cared for at the place where he'd worked for three decades quite moving. "I knew most of the people taking care of me," he recalls. "These are people I'd worked with most of my life. That felt terrific." Another surprise came four weeks later, after he returned to work. "My patients were thrilled," he says. "I'd see patients, and they'd say how proud they were of what I'd done."

By now, more surgeons were becoming interested in our ideas. Mike Rees was at this point collaborating with Steve Woodle, a veteran transplant surgeon at the University of Cincinnati who had received a liver transplant himself in 2003 after his own liver had been destroyed by cancer. The two men had started using software — an early version of which was written by Mike's dad — to identify compatible pairs. They asked Tayfun, Utku, and me to adapt our matching algorithms to their system, to help them figure out how to arrange exchanges to produce as many transplants as possible, subject to the criteria they used.

The following year, in January 2006, Steve asked me to give a talk — a Transplant Grand Rounds, as these talks are called — on kidney exchange at the University of Cincinnati College of Medicine. A two-way exchange that our software had identified was scheduled for that morning, and Steve invited me to watch. One set of surgeries was happening at his hospital, the other at Mike's hospital in Toledo.

That morning, Steve picked me up in his SUV, and we drove to the University of Cincinnati Medical Center and changed into scrubs. Two adjacent operating rooms were already busy with preparations. Simultaneously, the same preparations were under way in Toledo.

From time to time, Steve got on his cell phone and checked on the progress in Toledo. His final call confirmed that all four donors and recipients were fully anesthetized, with initial incisions made. Everyone was ready; no one had run into difficulty with anesthesia. Both nephrectomies were given the go-ahead.

The donor surgery I watched was a "hand-assisted laparoscopic nephrectomy." This procedure gives the donor a quicker recovery than the older surgery in which the kidney is removed through a much larger incision. The surgeon worked instead through two small incisions. Through one he inserted a camera and a light to project an image on a screen that let the surgeon see what he was doing and allowed the rest of us to watch. Through the other he inserted a cutting tool that resembled a tiny pair of scissors attached to the end of a knitting needle. The screen showed not only the instruments and patient's internal organs but also the gloved hand of an assisting surgeon, inserted through a larger incision. The two doctors worked as a team, with the seemingly disembodied hand responding to requests to put tension on tissues so they could be cut and cauterized. Like a rabbit being pulled from a magician's hat, the freed kidney emerged from the body in the second surgeon's gloved hand, to be immediately deposited in an ice-cooled steel bowl.

This bowl was carried quickly to the adjacent operating room and the waiting recipient. Before the kidney could be transplanted, it first had to be prepared. A textbook sketch of kidney anatomy shows blood entering through the renal artery and exiting through the renal vein. In reality, many smaller veins branch off the large central one, and these, too, have to be located and tied off. Steve and his colleague, the veteran surgeon Rino Munda, did this as a team, rapidly finding and tying off the many small blood vessels. The deftness with which they worked made me think of trout fishermen preparing fishing lures.

I had skipped breakfast, fearing that the sights and smells of the OR might make me sick, but that had been unnecessary. I was too fascinated to feel squeamish. Meanwhile, Steve and Rino were so relaxed that they had time to give me a running commentary as they worked. The main renal vein resembled wet tissue paper; it was hard to imagine it could be sewn at all. And yet Steve and Rino worked with practiced ease. They held their sewing needles with instruments that looked like giant tweezers and wielded them like extensions of their hands. When they gave a young surgical fellow the opportunity to sew a stitch in the artery (which is firmer and hence easier to sew than the vein), his awkwardness made it clear that their skills had taken years to master.

Paying It Forward

As surgeons and hospitals gained confidence and experience with exchanging kidneys, more ambitious types of exchanges started to become accepted. Now the idea that kidney exchange could be thought of as a mixture of cycles and chains seemed more practical than it had when we'd first proposed it. The most interesting kind of exchange chain sprang from the small but growing number of living potential donors interested in giving a kidney to anyone who needed one. Previously, such a non-directed, or "altruistic," donor gave his kidney to a patient on the deceased-donor waiting list. Now these donors were offered the chance to help save more than one life by starting a chain of transplants. The first link in such a chain would be the gift from the non-directed donor to a patient in the pool of patient-donor pairs, instead of to a patient on the waiting list.

John Robertson, from Portsmouth, New Hampshire, became such a donor in 2010 after he saw a story reported by CBS News. "It was about a woman in Phoenix who took a cab three days a week to dialysis," Robertson says. "She told the cabbie that she was going to die unless she received a kidney. And he said, 'You can have one of mine.'"

That tale inspired Robertson. He was semiretired—he'd sold his bookstore a few years before—and knew he could take time off for surgery and recovery. Still, he wondered if, at age sixty-two, he was too old to become a donor. A hometown hospital put him in touch with the transplant coordinator at Brigham and Women's Hospital in Boston. "I asked, 'Do you accept geezers' kidneys?' And she said, 'Yes, but you'll have to be awfully healthy.' The more I learned, the more I wanted to do it."

Robertson underwent the weeks of tests that gauged his health, as NEPKE searched for a recipient. "The toughest part was my impatience," he says.

While Robertson was getting restless, Jack Burns was growing desperate. He'd been diabetic since his thirties. Three decades later, his kidneys were failing. Without a transplant, he faced dialysis and the prospect of losing his job helping manage food services at Fenway Park. Dialysis is so enervating and time-consuming that many patients end up unemployed. His wife, Adele, wanted to give him one of her kidneys, but their blood types didn't match. Jack's transplant coordinator enrolled the couple in NEPKE. That May, the Burnses learned that they would be part of a three-way chain. A donor from New Hampshire—they weren't told his name—would give Jack a kidney, while Adele's kidney would go to someone on the waiting list. Surgery was set for June.

John Robertson's and Adele Burns's surgeries happened simultaneously. As soon as the surgeons at Brigham removed Robertson's kidney, they sent it around the corner to Beth Israel Deaconess Medical Center, where Jack was already prepped. Adele's kidney was implanted in a young man from Cambridge, Massachusetts. Where previously one donation would have occurred—from Robertson to a person on the waiting list—two happened instead. Some NEPKE chains at that time involved three donations and three transplants, or six surgeries total.

Why only six? Because NEPKE's chains continued to be done with simultaneous surgeries, multiple surgical teams and operating rooms had to be coordinated, which impeded longer chains. After so much success, this was a frustrating limitation.

In a 2006 paper, Tayfun, Utku, and I, together with Frank Delmonico and Susan Saidman, NEPKE's immunology specialist, had proposed that more transplants could result if the simultaneous-surgery requirement were relaxed. It proved to be a controversial proposal. To understand why it was nevertheless attractive, let's do a simple cost-benefit analysis to think about why conventional exchanges are done simultaneously.

With nonsimultaneous surgeries, in a conventional exchange between two pairs, a donor might renege and leave the potential recipient in the lurch. You can imagine how this would play out: I give a kidney to someone's brother today, in the expectation that my wife will receive one tomorrow. But when tomorrow comes, my wife's prospective donor backs out. I've given up my spare kidney—which means that we can no longer participate in some future kidney exchange—and my wife still needs a kidney. The broken link has irreparably harmed us. To prevent that kind of harm, exchanges in closed cycles are always done simultaneously.

But the presence of a non-directed donor can remove the risk of that kind of severe harm to a pair that gives a kidney and then doesn't get one in return. Now, every pair could be scheduled to receive a kidney before they gave one. And if the chain was broken unexpectedly—that is, if someone proved unwilling or unable to donate—no one would be irreparably harmed.

To see what I mean, let's add my wife and me to the Robertson-Burns chain. Imagine that we're in line behind Jack and Adele Burns: I'm scheduled to give a kidney after my wife receives one from Adele. The chain would start the same way, with John Robertson's altruistic donation to Jack. But since the surgeries aren't simultaneous, Adele has time to panic and back out. (The real Adele wouldn't have done that, but let's assume she did.) What happens? My wife and I are seriously disappointed, but we're no worse off than we were before Robertson came forward. I still have my kidney, and we can still enter into a future exchange. That reduces the costs of a broken chain and thus increases the attractiveness of allowing nonsimultaneous surgeries.

As noted earlier, when Tayfun, Utku, and I floated the idea of nonsimultaneous exchanges, we met with considerable resistance. Ruthanne Leishman at NEPKE told us that surgeons would never accept it. Frank, fiercely protective of the kidney exchanges he'd helped pioneer, worried about lawsuits if someone reneged on a promised donation—and that the resulting bad publicity might set back the whole program.

But in Ohio, Mike Rees was willing to take a chance. He'd already organized multistate exchanges through a nonprofit he'd founded called the Alliance for Paired Donation (APD). His first nonsimultaneous chain started with Matt Jones, an altruistic donor from Michigan. Jones was the manager of a National Car Rental office when he decided to give up one of his kidneys. Though only twenty-eight, he wanted to do something admirable that his kids would remember him by. Jones's donation in July 2007 set off a chain of ten transplants that stretched over the next eight months.

Jones started the ball rolling by flying to Phoenix and donating to a woman there. Her husband then donated to a woman in Toledo. By March 2008, the chain had passed through six transplant centers and five states. Twice, months stretched between the time the patient in a pair received a kidney and the time the donor in that pair gave one. Yet despite the long waits, no one reneged. In November 2009, *People* magazine declared Mike and the donors in that chain "heroes among us." And the chain hadn't ended yet: the last person in *People*'s display of twenty-one patients and donors was twenty-nine-year-old Heleena McKinney, the daughter of the last recipient. Under her photo was the caption "Donor-in-waiting." As it happens, McKinney was hard to match, but almost three years later a suitable recipient for her kidney was found, and she continued the chain, which eventually included sixteen transplants and ended when the last donor gave a kidney to a patient on the waiting list who didn't have a donor to continue the chain.

Thanks to Mike's chain and the publicity surrounding it, a revolution had begun. Potential donors realized that their gift might save ten lives, and more of them began contacting Mike's and other hospitals. Our report on that first nonsimultaneous chain in the prestigious *New England Journal of Medicine* gave it an imprimatur that allowed other transplant centers and kidney exchange networks to explore such chains with confidence. In the years since, nonsimultaneous chains have continued to grow, with scores of other hospitals and networks joining in.

One of Mike's most energetic disciples in making use of nonsimultaneous chains is not a surgeon but a businessman, Garet Hil, who learned about kidney exchange when his daughter's kidneys failed in 2007. Neither Garet nor several of his daughter's uncles' kidneys were compatible. Garet enrolled in as many kidney exchange programs as he could across the country and remembers gratefully the welcome he received from NEPKE and the APD. But his dealings with some hospitals running their own kidney exchange programs, including Johns Hopkins and the University of Pittsburgh Medical Center, frustrated him. "Several of them wouldn't let you in their program unless you flew to a distant city and went through the entire workup process with them," he recalls. "And not just me but my daughter, too, who was on dialysis. When they said, 'You have to move your daughter to our transplant center,' I said, 'My daughter's fine at Cornell's hospital [in New York City].' But they wouldn't budge." Garet concluded that these hospitals were putting their financial interests ahead of patients. He still believes they were "using paired exchange as a weapon to gain market share."

Within a few months, discussions over whether his daughter would have to travel for her transplant became moot. A cousin was able to donate, and the surgeries took place that July. A week after those surgeries, something happened that reaffirmed Garet's belief that patients should have their transplants close to home. His daughter's legs began to swell, and Garet and his wife feared that she was rejecting the new kidney. They hustled her to the emergency room at NewYork-Presbyterian, Cornell's hospital. "It wasn't a rejection," he says. "But if she'd been transplanted in a city a thousand miles away—what a nightmare that would've been. Your transplant center should be relatively close by."

Garet's frustrations led him in late 2007 to form an exchange network he called the National Kidney Registry. It is headquartered near his Long Island home and aims to facilitate potentially quite long nonsimultaneous chains by recruiting hospitals and non-directed donors. If a hospital sends a non-directed donor, the NKR promises to end one of its chains at that hospital. That ensures that the hospital doesn't "lose" a transplant by sharing its donor. Keep in mind that hospitals earn revenue on their transplants; they're commercial enterprises as well as caregivers.

Speaking of commerce, sometimes when I explain all of the market design, computer programming, and medical politicking that's required to bring about kidney exchanges, someone—sometimes a fellow economist—will tell me that buying and selling organs would simplify things. If we just let the market do its work, this person will say, the price of a kidney would settle at the place where enough people would be willing to sell their spares—and the waiting list would disappear. After all, it's not as if there's a shortage of highly motivated buyers.

As an economist, I understand that point of view. Markets often deliver what people want, and without the efforts of market designers. But the big obstacle to such buying and selling of kidneys is something I've already noted: it's illegal everywhere except Iran. That's because many people find organ sales repugnant. We'll explore repugnance as a constraint on otherwise plausible transactions in chapter 11. For now, it seems unlikely that the norm against selling kidneys for cash will change any time soon.

Hard and Easy Matches

In the meantime, well-designed exchange clearinghouses can help people who need kidneys. I don't think that kidney exchange by itself will ever fully eliminate the ever-lengthening kidney waiting list, but if we persuade enough patients, surgeons, and hospitals to participate and fully share their information, we can do many more transplants.

Yet as kidney exchange grows, it faces new obstacles, just as many markets do once they start to become established. When we began, the problem was to design kidney exchange in a way that would allow patient-donor pairs and their surgeons to enroll with confidence. Nowadays, the directors of transplant centers have become strategic players, and the biggest challenge is designing exchange clearinghouses in a way that makes it safe for hospitals to enroll *all* their patient-donor pairs, not just the ones that are hardest to match. At the moment, some hospitals are hanging on to their easy-to-match pairs so they can do their exchanges in-house. Only when such a hospital can't find an in-house exchange does it hand off a particular pair to a clearinghouse.

Withholding easy-to-match exchanges is a common temptation in markets with middlemen. Think about the market for real estate. When the market is hot, easy-to-sell homes may never appear on the market at all. Instead, real estate brokers will match sellers who list with them and who aren't asking too high a price with buyers who come to them for help buying a home and don't need to sell their own home before they can buy one. That might be good for those buyers and sellers, or it might not. It certainly makes for a quick sale, which saves some hassles, but the sellers might get a better price if they offered their home for sale more widely. There's no doubt that it's good for the real estate brokerage, however, since the firm captures both sides of the trade quickly, with little investment of time and effort. Notice that it also makes the whole market have to work harder, though, because it keeps the easy buyers and sellers out of the open market, leaving in the market a disproportionate share of high-priced houses and cash-strapped buyers. That may result in fewer overall sales, as some of the unaggressively priced houses might have been bought by buyers who are otherwise priced out of the market.

Stockbrokers face similar incentives, and so laws and industry regulations now exist to prevent brokerage firms from keeping easy trades in-house. It's tempting for brokers to save costs—and make additional profits—by buying directly from sellers asking low prices and then immediately selling to buyers offering high prices. Notice that this practice may be good for the brokerage firms but not necessarily for their customers, and certainly not for the market as a whole, which may be able to successfully serve fewer people than if the market were as thick as it would be if all trades were listed on financial exchanges.

In much the same way, when transplant centers withhold easy-to-match pairs and transplant them internally, it reduces the number of people who can be matched nationwide, because it's easier to find matches for hard-to-match pairs if they don't always have to be matched with other hard-to-match pairs. I've studied this problem with Itai Ashlagi, a professor at MIT, and we think the problem can be fixed with relatively small changes to existing practices, by extending a kind of accounting that already exists to track hospitals that initiate non-directed donor chains. Think of it as a kind of frequent-flier program for hospitals: We could track how many easy-to-match pairs each hospital enrolls. Then, whenever there was a tie between two hard-to-match pairs about which should be included in some exchange, it would be broken in favor of the pair with a patient from whichever hospital had enrolled easy-to-match-pairs, too. Unfortunately, that means explicitly recognizing that hospitals are strategic players in competition with one another—something everyone knows but many doctors and medical administrators find hard to acknowledge.

Meanwhile, the problem of hospitals withholding pairs is likely to worsen. That will make long nonsimultaneous chains even more important, because when hospitals withhold their easy-to-match pairs, the hard-to-match pairs who are enrolled will be unlikely to be able to engage in simple exchanges with just one or two other pairs, since it's hard to close the circle — hard to find that double or triple coincidence of wants — when each pair is hard to match.

Playing Well Together

The problems that are presently keeping kidney exchange from expanding to its full potential don't come as a complete surprise. Market design, after all, isn't just about understanding markets and figuring out how to organize them better. Politics is involved, too. I'm not thinking of "politics" as a dirty word, only that when you are dealing with a multibillion-dollar industry, like the care of kidney disease, there are institutional and career interests at play, and they react slowly and cautiously to new technological and organizational possibilities. Nowhere has this been clearer than in the efforts to organize kidney exchange at the national level.

Frank Delmonico's New England Program for Kidney Exchange and Mike Rees's Alliance for Paired Donation created thicker markets by bringing together patient-donor pairs from dozens of hospitals. Later, Garet Hil's National Kidney Registry joined this select group. But these three programs rarely manage to share their patient data to arrange transplants across networks. And many hospitals simply refuse to participate. So the market isn't as thick as it could be. This means some transplants are being missed.

It was clear from the outset that the best way to make the market thick enough to find all potential exchanges would be to organize a national kidney exchange. But two problems immediately presented themselves: one technical and computational, the other political and organizational.

The political and organizational issues have in many ways proved harder to address than the computational ones, and are at least as important for market design. The harsh reality is that hospitals have trouble coordinating with one another because they're also competing for patients. This has made it difficult for either the APD or the NKR to gracefully expand to a national scale, although both are making progress in that direction.

There is, in fact, a national organization with which transplant centers already communicate about deceased-donor organs: a federal contractor called the United Network for Organ Sharing (UNOS). But when Frank Delmonico became president-elect of UNOS in 2004, he discovered that the permanent staff there was not eager to take on the added responsibility of dealing with kidney exchange.

In 2010, UNOS began a national pilot program of its exchange clearinghouse. To date, the program has produced few transplants, although there are signs that situation may be starting to change. UNOS answers to many constituencies, which hinders its ability to move quickly to adopt the best practices developed by other kidney exchange networks. In the absence of an effective nationally organized exchange, clearinghouses like the APD and NKR may continue to expand and merge. But UNOS has had one bit of good news: NEPKE closed its doors at the end of 2011 in order to merge its operations with the national program, and NEPKE's clinical program manager, Ruthanne Leishman, moved over to UNOS. Prior to Ruthanne assuming her new responsibilities in the summer of 2011, the UNOS program had completed only two transplants. Following her arrival, it quickly performed another fifteen. But UNOS still has a very long way to go to catch up with kidney exchange as it grows throughout the country.

Regardless of whether we have a national exchange or a few big exchange networks, part of the work still needed to allow kidney exchange to continue to grow is that the federal government and insurers must figure out how to reimburse the costs of kidney exchange. The United States now finds itself in a bizarre situation in which Medicare and private insurers will pay for dialysis, which is more costly and less effective than transplantation, but won't fully fund all the work that makes exchanges possible. That's why hospitals sometimes rely on private pilots like my colleague Jerry Green, whom we met in chapter 1, to volunteer to carry kidneys from place to place.

I look forward to the day when kidney disease is just a memory, so that no kidney transplants will ever be needed again. But until then, I'd like to see as many patients as possible who need transplants get them. And while I'm often frustrated by how slowly we are making progress, I also never anticipated that we would make as much progress as we have in such a short time.

As I write this in 2014, kidney exchange has become a standard method of transplantation in the United States and is growing around the world. As experience accumulates, the evidence grows that potentially long nonsimultaneous chains are good for kidney patients, and *particularly* good for the hardest-to-match patients. Thousands of transplants have been accomplished that wouldn't otherwise have been possible. In recent years, the majority of these have been through chains.

Kidney exchange is very different from the markets we saw in chapter 2. But as I've tried to show, market design for kidney exchange is still about making the market thick, uncongested, safe and simple, and efficient. In the case of kidney exchange, making the market thick involved assembling databases of patient-donor pairs. Dealing with congestion initially had to do with being able to schedule enough operating rooms at the same time, and now has to do with organizing chains. Making the market safe and simple continues to involve making it easy for hospitals to enroll all of their patient-donor pairs, so that the market will efficiently help as many patients as possible get transplants.

For any market to work well, its marketplaces have to solve all those problems, although the solutions will be different for different markets.

Market design has another essential aspect, which has to do with human behavior. In recent years, behavioral economists have upended traditional economic assumptions by noticing that people aren't relentlessly calculating and purely self-interested, and market designers will miss good opportunities if they forget that. Consider non-directed kidney donors. If everyone acted purely out of self-interest (as old-school economic models sometimes suggest), altruistic donors wouldn't exist. And what about donors in nonsimultaneous chains? If most people cared only about themselves and their families and friends, more of them would renege and fail to donate their kidney after their loved one has received one—yet very few do.

Each stage in the design of kidney exchange has involved adjustments in the market design—a sort of dance between the mathematical models; the surgical logistics; and the patient, doctor, and hospital incentives, risks, and rewards. When we originally proposed that kidney exchange would integrate cycles and chains, we didn't anticipate that we'd have to start with simple two-way exchanges, or that when larger cycles and chains became possible, long nonsimultaneous chains would grow to play such an important role. Each of these developments involved a modification of the market design in response to changes in the conditions of the market and the behavior of the participants.

The general lesson to keep in mind as we look at more usual markets is that not only do marketplaces have to solve the problems of creating a thick market, managing congestion, and ensuring that participation is safe and simple, but *they also have to keep solving and re-solving these problems as markets evolve.*

And just as engineers learn a lot about how to build bridges by studying those that collapse, market designers can learn a lot about what makes markets succeed by studying those that fail. A bridge will collapse if its weakest part fails, and a market design won't succeed unless it avoids each of the ways that it could fail. Often the same competitive impulses that make well-designed markets succeed cause poorly designed marketplaces to fail.

In the next four chapters, we'll look at failures—of thickness, congestion, and safety and simplicity. Then we'll be in a better position to understand how some markets that were broken were able to be redesigned and repaired.

PART II

Thwarted Desires: How Marketplaces Fail

4

Too Soon

TO UNDERSTAND THE many ways in which markets fail, we must begin even before the beginning.

Part of making a market thick involves finding a time at which lots of people will participate at the same time. But gaming the system when the system is "first come, first served" can mean contriving to be earlier than your competitors.

That's why, for example, the recruitment of college freshmen to join fraternities and sororities is called "rush." Back in the late 1800s, fraternities were mostly social clubs for college seniors. But in an effort to get a little ahead of their competitors in recruiting, some started "rushing" to recruit earlier and earlier. Fast-forward to today, when it is first-semester students who are the targets of fraternity and sorority rush.

And that's not the only way the rush to transact sooner has entered the English language. It's also the reason that Oklahomans are called "Sooners."

The nickname was born on April 22, 1889, the beginning of the Oklahoma Land Rush, and truly entered the American vernacular four years later, on September 16, 1893, the height of the rush, known as the Cherokee Strip Land Run. In both cases, thousands of people — 50,000 in 1893 — lined up at the border of the former Indian Territory and, at the sound of a cannon shot, raced off to stake out free land.

At least that was the plan. And most participants abided by the rules — not least because the U.S. Cavalry was patrolling the Strip with orders to shoot anyone found in the open territory, or crossing the line, before the signal sounded. To prove the cavalry's seriousness, when one unfortunate soul — perhaps confused by a pistol shot — took off early, they rode him down and shot him dead — to the horror of thousands of onlookers.

When, finally, the cannon roared, those same thousands — on horseback, in wagons, and even in carriages — surged forward in the most famous photographic image of the era.

Fifteen miles away — in what would, by afternoon, be the bustling municipality of Enid, America's newest city — stood the only public building in the Strip, a land office/post office. About noon, the assistant postmaster, Pat Wilcox, took his binoculars and climbed up on the roof of the building. Looking south, he saw a lone rider, a twenty-two-year-old cowboy named Walter Cook, appear on the crest of a low hill. Tearing toward him, and then rushing on past, Cook jubilantly staked his claim to a plot of land at the very center of the planned city.

Cook had played by the rules, waiting until he heard the signal to take off. But lots of other people, despite the cavalry's draconian efforts, had crossed the line earlier. These "claim jumpers" would come to be called Sooners, for their timing — and in the long tradition of turning pirates, bank robbers, and other brazen criminals into lovable rogues, that would also become the nickname for all Oklahomans, and eventually for the University of Oklahoma's football team.

Gaming the system by entering the Oklahoma Territory to stake a claim before September 16 was illegal, but that didn't prevent it from happening. And claim jumping wasn't the only thing that didn't go according to plan in the course of that crazy day.

Take poor Walter Cook. His claim was quickly overrun by three hundred false claimants to the same plot of land, all taking advantage of the fact that the law wouldn't arrive for hours to validate anyone's claim. In the end, Cook got nothing except a lesson in the dangers of a poorly regulated, lawless market.

Cook might have had a chance if the land office had been open when he arrived and had processed his claim quickly. But instead, the line at the office quickly grew to hundreds of claimants, then thousands from throughout the Strip. Fights broke out; robberies occurred; at least one person died of a heart attack.

There were at least two ways in which the allocation of land failed to work well that day. First, the law-abiding citizens who followed the rules were often preceded by those who entered the territory sooner and marked their claims earlier. Second, the fact that those claims all had to be recorded at the land office in Enid on the same day led to congestion and confusion, in which even some of those who had arrived in time to stake a claim, such as Walter Cook, couldn't get it recorded. The market wasn't fast enough to deal with all the claims made that day, and so it couldn't always sort out which claims came first.

Sometimes the problems of going too soon are subtler. *Jumping the gun* — in Oklahoma it was literally a cannon — can cause potentially thick markets to unravel. They become thin when too many participants try to transact before their competitors are fully awake and present in the market.

Let's look at those other Sooners, the ones who play football for the University of Oklahoma. We turn to college bowl games to see how "too early" can ruin a matching market's ability to make good matches. The matching of football teams to play in the big end-of-season bowl games suffered for many years because the teams that would play in those games were chosen too soon to make for good matchups.

Make Me a Match, Catch Me a Catch

For those who love college football, there is no time of the year more exciting than bowl season, when the top teams from different conferences meet to determine which are better—and, ultimately, to determine the national champion. Unfortunately, most college football fans have come to believe—and sometimes argue vociferously—that the system is broken. And they're right.

For a long time, teams and bowls succumbed to the temptation to do deals early. And while college football isn't the most important market in the world (except to its fans), the fact that new information is available every weekend about which teams have won or lost, and that those teams are then ranked according to polls of sportswriters and coaches, shows very clearly how important information can be ignored when the market moves before the results of the final games of the season have been played.

As the television audiences and advertising revenues became important, bowl committees began to recruit teams to play in the bowl games earlier and earlier—indeed, so early that the teams they recruited were sometimes, after an unexpected loss or two, no longer championship candidates by the time the game was played. That's one of the dangers associated with early transactions: they can come well before important information is available. And that can mean bad matches made and good ones missed.

Which teams get to play in which bowls is handled differently today than it used to be. Sports fans can disagree about how well the current system works, but no one disagrees about how badly it used to work before the market was redesigned.

The bowls are independent businesses that control a stadium and make contracts with television networks and corporate sponsors. Each would like to host a postseason game between the two teams that are ranked best in the country at season's end. For many years, the National Collegiate Athletic Association tried to make bowls and teams wait long enough to get good matches. But it consistently failed to do so, and after the 1990–91 season, it gave up trying.

That season there were nineteen postseason bowls. The one that paid teams the most was the Rose Bowl, which was "closed": it had a long-term contract with the Big Ten and Pacific-10 football conferences, and each year the champions of those two conferences played each other in the Rose Bowl (and the two conferences shared the bowl revenues of their champions). So the Rose Bowl wasn't involved in the unraveling we're examining here; it merely waited until the conference champions were determined.

But the other bowls had different arrangements. The Fiesta Bowl faced a unique challenge: as an "open" bowl, it needed to find two teams to play. The other top bowls were all "semi-closed"—that is, they each had a contract with one football conference, whose champion would be one of the teams that would play. Meanwhile, each of these bowls needed to find one additional team to provide quality competition. The available pool consisted of teams that were not in any football conferences (independents) or were in conferences that were not contractually tied to any bowl.

In 1990, the NCAA rule was that teams and bowls couldn't finalize bowl matchups until "pick-'em day," which that year was Saturday, November 24. But some bowls and teams went ahead and made earlier arrangements. Notre Dame, an independent, had begun the season as the number 1 ranked team and had recovered from an early loss to regain that position by early November. Meanwhile, Colorado had overcome an early season loss to become the number 4 team in one of the rankings and number 3 in the other. When Colorado beat Oklahoma State and clinched the Big Eight conference championship, the university was ensured a berth in the Orange Bowl and rose in the rankings to number 2.

The next day, on Sunday, November 11, thirteen days before pick-'em day, an agreement was announced between the Orange Bowl and Notre Dame. This meant that the currently first- and second-ranked teams in the nation would meet in the Orange Bowl, and thus make that bowl the de facto national championship.

Announced the same day was Virginia's acceptance of a bid from the Sugar Bowl to play the still-to-be-determined Southeastern Conference champion. And following the Orange Bowl agreement, the University of Miami agreed to play in the Cotton Bowl against the still-to-be-determined Southwest Conference champion. At this point, Notre Dame, Virginia, and Miami all still had four games left in the regular season.

In college football, four games is forever. And sure enough, shortly after inking its agreement, Notre Dame lost a game and finished the regular season ranked number 5. Meanwhile, Virginia, which had lost only one game before its agreement with the Sugar Bowl, lost two games and finished the regular season unranked in one poll (meaning it wasn't even in the top 25) and number 23 in the other. In the end, no bowl succeeded in getting the number 1 and number 2 ranked teams (which turned out to be Colorado and Georgia Tech).

Thus, when the bowl games were over, there was no consensus national champion: Colorado was ranked number 1 in one poll and Georgia Tech in the other. Since they hadn't played each other, the sportswriters and coaches who were surveyed for the national ranking each felt entitled to his own opinion.

Faced with such a public failure to enforce pick-'em day, the NCAA abandoned the attempt for the 1991–92 season. The Football Bowl Association responded with an attempt to enforce a pick-'em day of its own and voted to levy a fine of $250,000 on any member that violated this understanding. However, the FBA was no more successful than the NCAA, and not surprisingly the 1991–92 bowls also failed to produce a matchup of the top two teams. Once again the postseason ended without producing a consensus national champion.

In retrospect, it's clear that several problematic market design features prevented good bowl matches. Because the Rose Bowl dealt with only two conferences, these conference champions risked not being ranked close to each other and would seldom be the two highest-ranked teams nationally. (But at least the Rose Bowl had a contract ensuring that the two teams that played each year would be the champs of their conferences.) The other major bowls enjoyed a substantial pool of conferences and teams from which to pick, but because of an unraveling of bids for their open slots, most were filled without the bowl committees knowing the end-of-season rankings of the teams invited to play. And because many bowls had one position reserved for a particular conference champion, this limited the matching flexibility of each of them, and of the market as a whole.

It wasn't simple self-restraint that stopped colleges and bowls from going early, nor could a powerful organization like the NCAA stop them. In the end, the unraveling didn't stop until the conferences and bowls figured out new rules that removed the incentives to determine bowl matchups before the final rankings were known.

They did this through a series of incremental, almost yearly reorganizations of the market, designed to make more teams available to be matched after the regular season — that is, to make the postseason market thicker. One way to do this was to enlarge the football conferences, so that the champions of each conference would be the best of a larger group of teams. By 2011, the Pacific-10 conference had become the Pacific-12 conference. The Big Ten kept its name but not its number: by 2011, it had also expanded to twelve teams, and then it expanded again to fourteen teams for the 2014–15 college football season. In addition, coalitions of bowls formed to make the market thicker, and eventually the Rose Bowl joined with the other major bowls to create the Bowl Championship Series (BCS) in 1998.

Now the number 1 and number 2 teams in the country, determined according to the BCS ranking system, played a national championship game, and this game was rotated from year to year among the participating bowls.

The fact that teams and bowls were matched later in the thicker BCS market doesn't necessarily prove that the market was working better. It's often hard to get a quantitative measurement of how well a matching market is performing in some ultimate sense—for instance, how much social welfare it is producing, beyond how well it is serving the participants in the market itself. But if we think of football games as entertainment, then how many people decide to watch the games isn't a bad measure of how well the market is working. When Guillaume Fréchette, Utku Ünver, and I looked at the Nielsen ratings for the televised bowl games over the years, we found that a game between the teams ranked 1 and 2 in the nation attracted so many more viewers that it was well worth it for the bowls to rotate such a game among themselves. This is why the BCS worked well when there was a consensus number 1 and number 2 at the end of the season and less well when there wasn't.

As I write this in 2014, plans are under way for a postseason playoff designed to more reliably produce a championship game that will attract many viewers. Four teams will be selected for the College Football Playoff, with the winners of the semifinals to meet for the championship game. The proposed new playoff model still has some of the old weaknesses of the BCS, but the uncertainty about which *four* teams to include should be less consequential in picking a national champ than uncertainty about which *two* teams to include.

The slow, incremental process by which the market for bowls evolved can be viewed as a kind of cultural evolution. Different bits of practice were reshaped over time, in ways that kept all the big players—the successful teams and conferences, the big bowls, the television networks—in business. Lots of interests had to be addressed to achieve any sort of coordination at all and to get some forward motion. Like football itself, that forward motion mostly came a yard at a time.

Incidentally, it's not just football *teams* that make early matches; often it's the players, too. For example, in 2012 Louisiana State University offered a football scholarship to Dylan Moses, a fourteen-year-old who had not yet begun the eighth grade and who wouldn't enter college for another five years. Who knows whether he'll be big enough, healthy enough, and accomplished enough to play when he is finally old enough for college. But coaches worry that all the other teams are recruiting early, and if they don't do the same, they could miss out on a future star.

This "sooner" mentality isn't limited to the more high-profile college sports. When I meet varsity athletes at Stanford University, where I work, I ask them when they first met a Stanford Cardinals coach. So far, the earliest answer I've gotten was from a player on the women's basketball team, who first met a Cardinals coach when she was in sixth grade. She hastened to add that she'd been a *very tall* sixth grader, on a team with older players, and that the coach had been surprised to hear that she was only in the sixth grade. He was scouting eighth graders . . .

Rush to Glory

Rushing to be sooner isn't just something in the history books or on the sports pages. If you know a recent college graduate who recently took a job with a big investment bank such as Goldman Sachs, there's a good chance that she'll get a call soon after beginning work. It will be from a big private equity firm such as Kohlberg Kravis Roberts, interested in signing her to a contract that would take effect after she's worked for Goldman for two years. And if you know someone who just graduated from law school and works for a big American law firm, he was likely hired by that firm initially as a summer associate, about two years before he earned his J.D.

Is this a good idea? Remember the 1991 Orange Bowl.

The same thing that happened then can happen to law firms that recruit years before their future employees earn their degrees. That top-notch first-year law student can go through a lot of changes over the next two years. At least the Orange Bowl selection committee knew how many teams – two – it would need for the game. Not so for law firms, which have to guess two years in advance how many lawyers they may need. Guess wrong, and they could be in a lot of trouble.

When a market's organization predictably leads to trouble, economists start asking whether it might be *inefficient,* meaning that a different organization might make everyone involved better off. We've already seen that going early can create bad matches, but it also could be that this approach benefits some people while hurting others. The market for new lawyers lets us see how unraveling can sometimes hurt *everyone.*

In particular, just about everyone could have been better off if that market had been less unraveled during the 2008 Great Recession, which reduced corporate demand for outside legal services. Hiring more than a year before the start of employment made it difficult for law firms to forecast their demand. Thus thousands of summer associates at large firms who'd accepted "permanent" offers shortly after their second-year summer associateships in August 2008 saw them rescinded or deferred before they started work in the autumn of 2009.

Some of the firms, to maintain their reputations and their relationships, paid those deferred employees a portion of their starting salaries and encouraged them to spend a year doing pro bono work – an outcome that was costly for both sides of the market.

If this two-year head start sounds bad, consider that in the late 1980s hiring was even earlier, with some students getting offers of summer associate positions right after they were *accepted* to a top law school, before they'd even taken their first class. Those firms undoubtedly would have liked to see how their prospective hires actually did in law school—but they worried that if they waited, other firms would snap up the best talent before them. So they told themselves that if Yale Law School wanted a student, that student also had a strong chance of becoming a good lawyer—just as at midseason Notre Dame had a good chance of being the number 1 team when it played in the Orange Bowl.

If making offers very early makes it hard to identify good job candidates, you might think that some firms would take a little more time and make offers to candidates who had already received at least one offer from another firm. But the firms that made early offers prevented this by making their offers *exploding*—that is, take-it-or-leave-it offers of such short duration that they didn't leave enough time for another firm to jump in and compete for the same candidate, or for a candidate to get another offer for comparison.

Exploding offers are common in unraveled markets. These offers are both early and short-lived. So not only are firms making offers before they have as much information as they'd like about how candidates might perform in school, but the candidates themselves are confronted with accepting or rejecting an offer before they know what other offers might become available. To put it another way, exploding offers make markets *thin* as well as early, and so participants are deprived of information about both the quality of matches and what kind of matches the market might offer.

In that situation, *nobody* has enough information to make an optimal decision.

More than the other sources of market failure that we'll explore, unraveling is a failure of *self-control.* Participants just can't stop themselves from transacting early, because if they resist the urge, they'll lose out to someone else. It's a little like what happened when my family planted a pear tree in our yard in Pittsburgh, right next to a wooded hillside. Each year, long before the pears were ripe, some squirrel would take them. I don't know whether squirrels like unripe pears or they just feared that if they waited any longer, the raccoons or the deer would get them.

Now, if a market is behaving badly and producing an inefficient outcome, it makes sense for participants to get together (if only for their own preservation) and design new rules to make the market work better. That's what happened in the 1980s. Student organizations, law schools, and law firms supported a rule-making organization called the National Association for Law Placement (NALP), which tried to bring some order to the lawless market for lawyers.

Because lawyers like precise rules, looking at these rules gives us a unique window on why unraveling is so hard to control.

One rule was meant to give brand-new law students a chance to learn a little law before being confronted with an exploding offer from a law firm. This rule said that if an offer was made to a student who hadn't yet completed the first year of law school, that offer had to remain open until the end of the first semester, in December.

Unfortunately, it's hard to make rules constraining lawyers, because many lawyers earn their living by obeying the letter of the law while evading its intent. So this rule worked for a year or two, until some lawyer, in charge of hiring for his firm, had the bright idea of writing an offer letter that said, essentially, *In keeping with NALP guidelines, this offer remains open until the end of the semester. But,* the letter continued, the job didn't come with much of a salary. There *was* a handsome signing bonus, however, which would bring the salary up to the usual level. *But* that signing bonus would be paid only if the offer was accepted *immediately.*

Regulating the market for new lawyers soon became an arms race between the rule makers and the rule breakers. As of this writing, the most recent NALP rules say that exploding *bonuses* are against the rules, too.

Rush to Judgment

At least lawyers and law firms make a show of obeying the rules while seeking ways around them. In the most prestigious part of the market for young lawyers, that of federal appellate judges hiring top students as law clerks, many judges openly flout the rules. Or perhaps a more "judicious" way to put it is that federal judges think they can make up their own rules.

Clerking for an appellate court judge is just about the classiest first job an ambitious young lawyer can have. For one thing, it's a ticket to the kind of career that makes people want to be lawyers. That's one reason the obituary of a retired senior partner of a big law firm often mentions his clerkship of decades before. (The first sentence might read: "Clancy Goldfinger, former managing partner of Catchum, Killum, and Eatum, who graduated from Harvard Law School in 1951 and clerked for Judge XXX and Justice YYY, passed away Tuesday.")

So there's a lot of competition among the best students at the elite law schools to clerk for one of the relatively few federal appellate judges. But at first glance, the clerkship market doesn't look like one that should experience unraveling, although it's easy to see why a law student would be tempted to accept an early offer from an appellate judge. Since there are so few judges and so many law students, however, every judge could get a very well-qualified clerk if only he or she would wait to see which law students did well.

But while there are only a small number of appellate judges, those judges realize there are an even smaller number of law students who will win the top awards at their schools or be elected to edit their law reviews. And those appellate judges are organized into circuit courts, not all of which are equally prestigious. Neither are all judges within a given circuit equally likely to have their clerks move up to the U.S. Supreme Court for a second, even more prestigious clerkship there.

So if all judges waited to recruit only third-year students as clerks—when it is clear who will be a law review editor or top student—only the most prestigious judges would be able to hire the best students from the handful of elite law schools. That's a very good motivation for slightly less prestigious judges to make offers before the students' third year.

It takes a brave student to turn down an offer from, say, a judge in the Ninth Circuit Court of Appeals (which covers all of California and more) in the hope that if she waits, she might get an offer from the even more prestigious D.C. Circuit. That could happen if she's lucky. But if she's only a little less lucky, she may have to settle for a much less attractive job than the one she's just been offered—and that she has to accept immediately or not at all. Of course, the judge is gambling, too: a student who looks likely to win law school honors may fail to do so, and may turn into a clerk who won't live up to her early promise. If the market ran later, the matching of students and judges would be more predictable, with the top jobs reliably going to students who had earned the top honors.

Notice that the law students who get these early offers are hardly facing the prospect of unemployment. But that doesn't mean that they aren't facing difficult decisions. There will be positions for them even if they wait, but maybe not such good positions. They have to make quick, strategic decisions taking into account what the rest of the market is doing.

Wedding Bells' Toll

Few of us will ever get an offer to clerk in a federal court of appeals. But once you understand this kind of strategic decision making, you'll begin to see it all around you, from marrying to finding a parking spot. Quite a few of us may face such a dilemma when deciding whether to marry a current girlfriend or boyfriend, or to break up in the hope of finding a better match later. That's a different decision when the market is thick, such as when you're in college and there are lots of single people your age, than when the market is thin, such as when most of the people your age are already married. And some marriage markets are tougher than others. Consider the teenage Bedouin bride in whose community polygamous marriages are common. Just such a woman lamented, "If you are 20 or older, you may be married as a second wife."

But even teenage brides don't face the earliest marriage decisions in the world. In some times and places, the marriage market has unraveled to the point that newborns are betrothed. In developing countries, it isn't unusual to find marriages arranged quite early, particularly for women, and particularly in places where women are in short supply because men compete for multiple wives. Some countries, such as India, have tried to stop this practice with minimum marriage age laws. But those laws have proved difficult to enforce, because private and informal matchmaking arrangements have emerged.

In searching for a striking example of unraveling, Xiaolin Xing and I considered places where child marriages occur, and even primitive societies in which unborn children may be betrothed. The most striking example we found involved an aboriginal people of Australia, the Arunta. Because the Arunta were polygamous, there was a relative shortage of women.

Marriages among the Arunta were frequently arranged between two men, one of whom had just fathered a baby boy and the other a baby girl. When two such men met to arrange a marriage, however, the union they were arranging wasn't between those two babies — it was much too late for that, because the baby girl's marriage had already been arranged. Rather, the two fathers were agreeing that the baby boy would marry the first daughter of the baby girl. That is, they were agreeing that the infant girl would become the *mother-in-law* of the infant boy. This was a marriage arranged by the father of the infant boy on behalf of his son and the father of the infant girl on behalf of his *granddaughter* by his infant daughter. In Arunta society, marriages could be transacted more than a generation in advance of when they would be consummated. You can understand how, as a responsible young father, you wouldn't feel safe letting your son's – or your granddaughter's – marriage arrangements lag behind their competitors'.

Notice that in many developed countries, ages at first marriage are increasing, not decreasing. As more women seek higher education and professional careers, they wait to get married. When I say it that way, I'm focusing on the choices made by women. But a woman can't simply choose a spouse, and neither is the choice of when to get married entirely an individual decision for either men or women.

Think back to the days when few women went to college. In 1947, for example, there were more than twice as many men as women in American colleges. A lot of people eventually married their high school sweethearts, because high school provided a thick marriage market in which one could find a lot of single people of the opposite sex, and those opportunities wouldn't be so abundant later.

By 1980, many more men and women went to college, and in equal numbers, so there were opportunities to make a match there, too, and the pressure to marry early was reduced. Today the growth of Internet dating sites also offers the possibility of a thicker marriage market for college graduates. Postponing marriage when there is still a thick market in the future isn't so risky, and more-mature brides and grooms might have a better chance of recognizing a good match.

So the timing of transactions depends not just on what is available now, but what is likely to be available later. When you're driving down a crowded street hoping to find on-street parking, you regularly face, for lower stakes, a decision like the one confronting a law school student with an exploding offer or someone thinking about marrying his high school sweetheart. You see a vacant spot while you're still some distance from your destination. Should you take it? This spot probably will be taken before you can loop around and return, and you may have to settle for parking even farther away or in an expensive garage. Or should you risk waiting for a better choice: a spot right in front of your destination? That would be a safer choice if you knew that lots of parking was available near your destination.

As you can see, unraveled markets aren't hard to find. We've just seen unraveling in matching markets ranging from sports to law to marriage, and in simple choices from whether to wait until a pear is ripe before picking it to how far from your destination to grab a parking spot.

But this kind of market failure wasn't widely remarked on when I first noticed unraveling in the 1980s, while studying the market for new doctors. Back in the 1940s, medical school students had to line up their first positions two years before they were due to graduate from medical school. Looking at it from the reverse perspective, hospitals had to hire their new interns and residents from a pool of medical students who hadn't even started on the clinical portions of their medical school education. Each side felt, correctly given the circumstances, that if they didn't move quickly, the good positions and the top students would already be snapped up. It was a mess.

I first thought that unraveling was unusual, a kind of rare accident that happened in relatively special markets, such as that for doctors. But as we've seen, lots of markets unravel. In fact, it's an even more widespread phenomenon than I've suggested thus far. For example, many selective colleges now fill more than half of their freshman slots through "binding early admission," a kind of exploding offer in which students apply early and commit to attend that school if accepted, without applying to other colleges.

Meanwhile, some private schools even enroll students at birth. At the Wetherby School in England, a school Princes William and Harry attended, the spaces reserved by newborns fill up early each month, and the school advises women scheduling cesarean sections to have them on the first of the month, if possible, to get a place before all the spots are gone.

In fact, unraveling is an ancient problem. In medieval England, it was sometimes a crime, called "forestalling," to trade before the official opening time of a market. It's not a crime today, but try telling that to the vendors who come to the farmers' market near my home and refuse to sell to me if I show up before the official opening time, for fear of starting a race with the other sellers about who can set up first.

Those farmers are exerting self-control, maybe with a little help from the city, which licenses them to use the street only between certain hours. But unraveling can't always be limited by self-control. Even if a law firm can control itself and delay its own hiring until some reasonable time, if the firm's competitors hire earlier, it could be caught short. This fear of being left behind is part of what turned pioneers into Sooners.

So unraveling is hard to control in a market that relies on self-control. Even if *you* have a lot of self-control, all you need is the suspicion that other participants might jump the gun and you will do so, too. It would be irrational not to. In many markets, what we see at first is a slow unraveling that suddenly tips over into a mad dash. It is often only then, when the participants see the profits from going a little early now getting swallowed up by the costs of racing with everyone else to go very early, that a consensus finally grows in support of reversing this unraveling. That's when new market designs might be considered.

Let me tell you about a simple market design solution that halted and reversed unraveling in a market in which unraveling was the only market failure that remained to be solved. The trick was to remove the need for self-control among those tempted to make early offers by handing over some control to the people on the receiving end of the offers.

Finding the Guts to Wait

If you're under fifty, you probably don't have to know what gastroenterologists do. Let's just say they are doctors who look after your digestive system, and after you turn fifty, you're supposed to visit them so they can look for early signs of colon cancer.

To become a gastroenterologist, a doctor must participate in what's called a *fellowship*, which takes place following his first job, or residency, after graduating from medical school. The market for medical residencies was the first unraveled market I studied. Today that market is no longer unraveled, and new doctors are matched to residencies during their last year of medical school, in a market that is thick, uncongested, and safe. (I'll tell that story in chapter 8.)

The medical residency that future gastroenterologists must complete is in the field of internal medicine and takes three years. So gastroenterology fellows could, in theory, be hired after they have three years of medical experience. Unfortunately, the unraveling of the fellowship market caused that hiring to creep back earlier and earlier, until first-year residents sometimes found themselves being interviewed for jobs that they might begin two years in the future. Once again, this could be costly, both for fellowship directors hiring fellows while they are still inexperienced, and for young doctors having to choose a subspecialty before they've had time to learn what they like.

When my colleague Muriel Niederle and I studied the unraveling of this market, we observed that fellowship directors were increasingly hiring applicants who had done their residencies locally, because the only first-year residents it was safe to hire were those for whom those directors could get reliable recommendations from their own colleagues.

This restriction in the candidate pool reduced the desirable diversity of fellows. What these directors didn't appreciate—until they saw our results—was that this local hiring was happening to everyone. Only then did they all realize that their own problem was in fact market-wide. As you might imagine, that generated a lot of interest in hiring later.

Muriel and I eventually helped them to plan a clearinghouse that operated later in the careers of medical residents, like the one that matched new doctors to residencies. But those same fellowship directors didn't trust each other to cooperate and wait for the clearinghouse; they all worried that the others would continue to hire via early exploding offers. If they waited to take part in the clearinghouse, they feared all the best candidates would already be hired.

This lack of trust threatened to keep everyone making early offers, just in case everyone else did—even when no one, or almost no one, wanted to. So we asked the four principal professional organizations of gastroenterologists if they couldn't simply forbid their members to hire before the clearinghouse opened for business. They told us they had no power to regulate the behavior of their members, the fellowship program directors.

We next asked those organizations if they could pass a resolution that would empower fellowship applicants who had accepted very early offers to change their minds if, later, at the time of the clearinghouse, they regretted their early decisions. That proposal caused some concern: administrators worried that the market would have many offers accepted and then rejected. Using several kinds of evidence, we were able to convince them that this wouldn't happen, since the incentive to make an offer before you could tell which applicants were good ones would be eliminated if early offers and acceptances weren't binding. By freeing fellows to change their minds if they accepted an early offer, the new approach deprived program directors of the incentive to make early offers and relieved them of the fear that others would do so. Thus they could safely wait and match to a great candidate later when the clearinghouse opened.

Part of our evidence came from the market for new Ph.D. students in universities. Almost all American universities have agreed that students shouldn't have to accept their offers before April 15 of each year. If students are pressed to accept an offer before that deadline, they can accept and then later decline, in order to accept another offer before that date. This single rule has virtually eliminated all exploding offers for Ph.D. candidates in the United States.

Another part of our evidence was experimental: when we set up these rules in the laboratory and ran them in a simple artificial market, they eliminated exploding offers.

Still another part of our evidence was theoretical. Exploding offers won't occur when everyone has enough experience with the market to know what to expect. When that happens, economists say the market is "at equilibrium." In this case, equilibrium meant that everyone would expect that fellowship programs would be committed to hiring the young doctors to whom they made early offers but who subsequently performed *below* expectations. But they wouldn't get to actually employ those accepting early offers who *exceeded* expectations, because those people would take better offers later. Since the main point of early exploding offers is to "capture" better candidates than you could if you waited, program directors wouldn't make exploding offers if they no longer accomplished that goal. It doesn't take a lot of self-control to stop making early offers if they no longer get you what you want.

This worked for the gastroenterologists: they accepted the arguments and implemented the advice, then successfully organized a clearinghouse that now operates each year much closer to the time that gastroenterology fellows will actually start work. Exploding offers aren't a problem anymore, and so almost everyone succeeds in hiring and being hired in the clearinghouse, which operates later, along the lines of the successful market for residents. That clearinghouse provides a thick market that is worth waiting for, much as a thick marriage market in college makes it less pressing for people to marry their childhood sweethearts, or a lot of parking spaces near where you want to park makes it easy to pass up parking spots that you encounter when you're still far from your destination.

Our solution to the problem of hiring new gastroenterologists highlights one of the crucial facts about market design: successful designs depend greatly on the details of the market, including the culture and psychology of the participants. In the years that followed, we encountered a number of other markets facing problems that at first glance looked identical to the gastroenterology market. But in the end, some of these markets required very different solutions.

Cultural Shift

One good example of this is the market for orthopedic surgeons —which at first seemed to be a near clone of the market for gastroenterologists.

When I spoke with the orthopedic surgeons at Massachusetts General Hospital, it quickly became apparent that they had an unraveling problem: they were hiring new fellows up to three years ahead of time, when the fellows were still young surgical residents. The senior surgeons weren't too worried that they couldn't assess the dexterity of these residents while they were still so young, but they'd noticed that some of their new hires, when they eventually showed up to take their positions, had matured into operating room bullies, who made the nurses and others reluctant to work with them. This complicated scheduling and was bad for morale. If the senior surgeons could wait until after the young residents had time to grow into chief residents and assume more responsibility, they would be better able to assess what kind of colleagues the new surgeons would be and not just how good they were with their tools.

When Muriel and I began to look into the details, we found that the hiring of orthopedic surgeons looked almost exactly like that of gastroenterologists—early exploding offers, local hires, and all. So, naturally, we suggested to them that the solution that worked for the gastroenterology market might work for them as well: that is, if they could empower applicants to change their minds after accepting early offers, those early offers would cease, and an orderly clearinghouse could be organized at a convenient later date.

But the orthopedic professional organizations — which include at least nine distinct subspecialties — quickly told us that they couldn't empower young surgeons to change their minds about agreements they'd made with senior surgeons. That would never happen, they said: senior surgeons were too powerful and imposing for younger surgeons to feel that they could really change their minds, no matter what anyone said. They saw no obstacle, however, to imposing sanctions on fellowship directors who made early offers. One of the professional societies even told us they simply wouldn't let those doctors present papers at their annual meeting. So, by way of blunter methods, orthopedic surgeons were also prevailed upon to stop making exploding offers, which allowed some clearinghouses to be organized in orthopedic subspecialties.

Orthopedic surgeons needed a somewhat different market design than gastroenterologists to fix similar market failures because the two professions had distinctly different cultures. But in each case, they were able to find a way to prevent exploding offers.

The problems facing federal judges in the clerkship market are harder to solve, because that market culture actually combines the difficulties faced by gastroenterologists and orthopedic surgeons. The organizations of judges — called "judicial conferences" — are like the gastroenterology organizations in that they have no way to prevent judges from making early exploding offers or punish those who do. Meanwhile, as with junior and senior orthopedic surgeons, law students aren't in a position to break promises to federal judges. These things make it difficult for judges to organize themselves in a way that lets them trust one another to obey the rules.

Markets unravel despite the collective benefit of having a thick market in which lots of people are present at the same time, with many opportunities to be considered and compared. Without a good market design, individual participants may still find it profitable to go a little early and engage in a kind of claim jumping. *That's why self-control is not a solution:* you can control only yourself, and if others jump ahead of you, it might be in your self-interest to respond in kind. These early movers become the equivalent of the Sooners in the Oklahoma Land Rush.

For both gastroenterologists and orthopedic surgeons, success had to do not just with setting a particular time at which the market should operate but also with having a well-designed market available at that time. As we'll see in the next chapter, making the market operate within a narrow time frame—but without providing something, such as a clearinghouse, that brings order to the market at that time—usually isn't a good enough solution to the problem of unraveling. It can cause congestion, as when members of an unruly crowd all try to stake their claims at the same time—which can result in a different kind of market failure, when people feel pressed to make offers (and demand replies) too fast rather than too soon. A congested market may break down in a way that makes the participants risk the fate of poor Walter Cook, the man who was soon enough to stake a claim but not fast enough to register it—the biggest winner, and the biggest loser, in the Oklahoma Land Rush.

5

......

Too Fast: The Greed for Speed

BEING *TOO EARLY* isn't the only way speed can prevent markets from achieving the thickness they need to succeed. Markets can also move *too fast.*

Speed can make a market thrive or tear it apart. Speed helps participants in a thick market to evaluate and process lots of potential transactions quickly. But sometimes making markets work faster also makes them work worse.

We've seen how speed—in the sense of starting things too early—can cause potentially thick markets to unravel, because not everyone likes everything about a thick market. Buyers typically like to choose among many sellers, and sellers like to see lots of buyers. But those same buyers don't want a bunch of other eager buyers driving up prices, and sellers hate that competitors might deprive them of a sale. Those attitudes don't disappear in a thick market.

The rush to move a little faster—and not just earlier—than one's competitors has changed many markets, from finance to sports, as well as many labor markets, including law and medicine. Let's take a look at each in turn, starting with the fastest market of all: finance.

A Game of (Milli)Seconds

Not far from where wheat futures are traded at the Chicago Board of Trade, there's another marketplace, the Chicago Mercantile Exchange. And near them both, at the University of Chicago, an innovative market designer named Eric Budish (a former student of mine) has been thinking about high-speed trading at both exchanges.

Budish is looking at the growing use of computerized algorithms and how they influence financial markets. He's also looking at how changes in the design of financial marketplaces might help solve some of their long-standing structural problems.

The Chicago Mercantile Exchange is a lot like the New York Stock Exchange. For one thing, they both have similar market designs: trades are made via a continuous electronic limit order book, which records the offers to buy (*bids*) and the offers to sell (*asks*) starting with the highest bid first and the lowest ask first. Anyone can sell or buy at any time by accepting the best bid or ask for *x* units of the financial commodity being traded.

The CME and NYSE also offer similar financial commodities. For example, in both marketplaces you can invest in the whole bundle of companies that make up the Standard & Poor's 500 stock index, which serves as a proxy for the overall American stock market. On the NYSE, you can conveniently make an investment by buying shares in an S&P 500 exchange-traded fund, listed as SPY. On the CME, you can take a similar position by buying E-mini S&P 500 futures, listed as ES. The price of SPY on the NYSE and the price of ES on the CME are very highly correlated: they move up and down together—not surprisingly, since they're both investments in the same bundle of companies.

The markets for SPY and ES are thick. Millions of futures contracts and exchange-traded shares are sold every day. At any hour when the markets are open—even at any minute—buyers can find many sellers, and sellers can find many buyers, so the price competition is intense.

But the situation changes if we look not at hours or minutes—or even seconds—but at thousandths of seconds, *milliseconds.* (There are 86 million milliseconds in a day, and it takes well over a hundred to blink an eye.) Even in the ultrafast world of finance, many milliseconds can pass with no trades taking place at all.

Thus a market that looks thick on a human timescale, with hundreds of opportunities to trade in the course of a single second, can look comparatively thin to a computer. Millisecond after millisecond ticks past, with no transactions and no change in the highest bid or lowest ask.

Just as important, it takes a few milliseconds for news of a price change in Chicago to reach traders in New York, and vice versa. Put another way, when the price of SPY or ES changes in one city, there can be a few milliseconds of lag time in which someone who has heard the news can buy cheap on one market and sell at a higher price on the other.

How fast do you have to be to make money this way? Before 2010, market news between Chicago and New York was transmitted fastest on cables that ran along the rights-of-way of roads and railways. But that year, a company called Spread Networks spent hundreds of millions of dollars to build a high-speed fiber-optic cable that went in a much straighter line and cut round-trip transmission of information and orders from 16 milliseconds to just 13. That 3-millisecond differential basically meant that only traders who used the new cable could make a profit by trading on momentary price differences between Chicago and New York.

Because an electronic order book—an important element of the markets' current design—is "first come, first served," whoever trades first gets the deal. And there are big profits to be made that way. So it won't shock you to learn that since 2010, billions of dollars have been spent on even faster cables. As I write this in mid-2014, the fastest price quotes are now carried through a microwave channel with a round-trip time between the two cities of just 8.1 milliseconds.

Now, it may not be a problem that faster traders can make more money, but it certainly isn't good news that billions of dollars are being spent on faster cables that don't make the markets work any better or provide any other social benefit. In fact, faster cables make the market work worse, and in ways that harm other traders.

Here's why. One way financial markets provide thickness to ordinary traders is by giving certain professional traders the incentive to become "liquidity providers." Members of this group don't plan to hold a security and watch it appreciate; rather, they move fast and act as market makers. They always offer both a bid and an ask on the financial commodity for which they are making a market; that is, they simultaneously offer to buy and sell. Liquidity providers make their money by having a spread between those buy and sell offers, which they continuously adjust as the market moves. The narrower that spread, and the greater the quantity they offer to buy or sell at that spread, the better service they are able to provide to whoever comes on the market to make a trade (and the more likely *they* are to actually get the trade, instead of their competitors).

But when there are high-speed traders in the markets, liquidity providers are forced to quote bigger spreads or offer to trade a lower quantity to at least partially protect themselves against being "sniped" by a trader using one of the new superfast lines. Such a trader might be able to buy from them at their old prices (now out of date, or "stale") and then moments later sell back to them at the new higher prices. The wider the spreads the liquidity providers quote, the further prices have to jump before they can be exploited on both sides of the trade this way, and the more they pass on the cost of protecting themselves to ordinary investors.

Very high speed trading can also contribute to instability in the market. A famous example, in which high-speed trading of ES futures and SPY exchange-traded funds was implicated, is the "flash crash" of 2010. In just four minutes, the prices of futures and of the related SPY exchange-traded funds (as well as many of the stocks in the index) were driven down by several percentage points—a very big move, in the absence of earth-shattering news—and then recovered almost as fast.

A subsequent investigation by the Securities and Exchange Commission and the Commodity Futures Trading Commission suggested that this brief distortion resulted from high-speed computer algorithms trading with one another, at a speed that eluded human supervision, and briefly spun out of control before anyone could react.

In the aftermath of this flash crash, there was added confusion involving order backlogs and incorrect time stamps that made it difficult to determine which trades had actually gone through, since even some of the market computers had been left behind by the high-speed traders.

Of course, fast cables and computers aren't the only ways to get information about the market and to act on it before others do. There are well-established laws against insider trading, which forbid company executives or others with privileged information to trade on, or share, that knowledge with, say, interested hedge fund managers before that information is publicly released. Those public announcements typically happen after the markets close, so that traders will have time to absorb the information before having to trade again. But America's prisons also house a sizable collection of high-powered financial movers and shakers who couldn't wait, some of them put away by the New York State attorney general, whose territory covers Wall Street.

The attorney general is involved because laws against insider trading are meant to level the playing field and make it safe for ordinary investors to buy and sell stocks. There's a similar concern when competition based on speed displaces competition based on price. That kind of arms race, which is more than a little like insider trading (because those who benefit from it are using knowledge that is not yet available to all), is bad for stock markets. Price competition among well-informed traders is one of the things that keep markets healthy.

In response to this growing crisis, Eric Budish and his coauthors, Peter Cramton and John Shim, suggested a simple change in the design of these financial markets that would restore price competition —and make it unnecessary to try to get information milliseconds before other traders. They proposed that instead of running a continuous market in which the first trader to take a bid or ask gets the trade, these same markets run only once per second.

In this proposed market model, during the 1 second between markets (a seemingly endless 1,000 milliseconds!), offers would be accumulated—and trades would be conducted at the price at which supply equals demand—among the traders who have offered to trade at that price. So instead of trades going to the fastest traders, they would go to those who had offered the highest bids and the lowest asks.

In such a once-per-second market, no one would have to wait very long on a human timescale: economic news that changes prices can wait a second if it must without disrupting the efficiency of markets. This type of market would also make it easier to keep track of trades: off-the-shelf computers would be quite sufficient for the task.

Mike Ostrovsky, a Stanford professor and market designer (who is also a former student of mine), tells a story that I've adapted here to make it easy to see the difference between "competition by speed" and "competition by price."

Suppose there was a football field in Chicago, on the flight path to and from O'Hare International Airport, and one day every arriving and departing plane started dropping money on this field—a billion dollars per year—that anyone who was fast enough could pick up and keep. What would happen? Well, that would quickly become a *very* popular football field as mobs of people began to show up and race for the falling bills.

Needless to say, a billion dollars a year is a lot of money, and it wouldn't be long before firms began to hire fast runners. Meanwhile, other firms would invest in machines, and soon faster drones would be snatching falling bills from the air just a little ahead of other, slower drones.

How much money would competitors find worthwhile to invest in this race for ever-faster dollars? Probably a very sizable fraction of that billion dollars. And unless there were a lot of unexpected other uses for fast bill-snatching drones, most of that investment would be a waste of resources as far as society was concerned, although it might bring handsome returns to the fastest bill snatchers of the moment.

Now suppose that instead of a high-speed race, the airport authorities sealed off that field, collected all the bills that fell there, and sold them at auction at the end of each day. There would still be competition, but of a different form. Now the daily haul would go to the highest bidder. Since a dollar is a bargain when it sells for only fifty cents, pretty soon competition would drive the prices nearer the value of the bills being auctioned each day. And over the course of a year, a billion dollars in bills would likely be sold for something close to a billion dollars. You might even call that the fair market price—it certainly would be the competitive price.

Note that, unlike buying drones and hiring Olympic-level sprinters, submitting bids doesn't take a lot of resources. Although firms might still invest in getting an accurate estimate of how many dollars fell each day, there would be a limit to how valuable it would be to know whether that amount was $2,739,727 or just $2,739,726. That's because the firm that is better at making that estimate can't earn a billion dollars a year; indeed, it might be able to earn just a few bucks' profit each day, which might not even pay the salary of a full-time bidder.

In this scenario, when bidders compete by price rather than speed, the yearly haul costs about what it's worth, so there's very little temptation to waste vast sums on otherwise unproductive competition. And if the money collected in this way were to be refunded to the passengers from whose luggage the falling money came, the market could also reduce the cost of air travel—a social benefit—rather than enriching the fastest traders.

It's hard to say what the chance is that the sensible proposal for once-per-second markets put forth by Eric Budish and his colleagues will soon be adopted so that the greed for millisecond speed can be replaced by once-per-second price competition. Financial markets are regulated, so many things can happen quickly if an idea is adopted by regulators. (And this idea *has* been endorsed by the New York State attorney general.) But in the absence of sufficient pressure by regulators, a brand-new market design is seldom adopted before a market becomes so dysfunctional that its users grow desperate for something new (or until an entrepreneurial market maker sees a way to compete with existing markets by offering a better design). It's not clear whether the financial markets have reached that state of dysfunction yet.

As the tale of these financial markets makes clear, a superior market design isn't always implemented. Building a better mousetrap isn't always rewarded when the mice have a say in the matter.

Financial markets are part of an enormous industry. The current winners in the race for speed were simply responding to the extant market design. They wouldn't be happy if their big investments in faster microwave channels were rendered useless. Yet they already know that could happen at any moment by the construction of a newer and faster communication channel. So a rule change leveling the playing field wouldn't be that much of a shock, and if it were phased in over a period of time, it might even earn their support.

For example, if it were announced that once-per-second markets would be established within a year, that would immediately discourage big investments in speed and allow today's fastest traders to enjoy their advantage a bit longer before the change. I mention this because part of market design involves recognizing that good ideas may not be enough on their own to fix a market. It's often also necessary to gather broad support from participants to get those ideas adopted and implemented.

So it isn't just a matter of good guys and bad guys. The interests of a wide range of participants must be taken into account to make sure that a new market design benefits as many people as possible.

The Victorian Internet

It might be good for the financial markets to slow down to human speed, but it wouldn't be good at all to slow them down so much that they couldn't react promptly to changes in the world and in the demand for trades. That's one reason that more and more financial marketplaces use computers to process trades themselves. But if we go further back in history, we can see how speeding up the movement of information can have a beneficial effect on markets—especially if that speed is measured in days rather than in milliseconds.

In the nineteenth century, the cotton market was one of the world's biggest markets, and the United States was a major player in it. America produced cotton, and English mills turned it into fabric and finished products made from cotton. When the first transatlantic cable was completed in 1858, messages between England and America could be sent by telegraph instead of by ship—in hours rather than days. The telegraph has been famously described by the journalist Tom Standage as "the Victorian Internet." Just as the Internet has resulted in an information revolution in our time, so too did the telegraph cause a revolution in the years surrounding the American Civil War.

Before the transatlantic cable, it took about ten days for a ship to bring price information from England to New York, and another ten days for a ship filled with cotton to make the return trip to Liverpool. Thus responding to prices on the English market (which reflected supply and demand there) essentially took the better part of a month.

After the cable was completed, price information could cross the Atlantic in a day. As a result, cotton shipments began to more closely match the fluctuations in the market, and prices on the cotton market became less volatile. Before the undersea cable, it was harder to match supply with demand because the news that arrived in New York was already more than a week old—slower than the pace of events even in that era. Speeding up the news let traders react to the market better. Speed had made for a better market in cotton.

But as we'll see next, speeding up the arrival of news so that decisions can be made with better information is different from speeding up decisions so that they have to be made before news arrives.

Let's take a quick look again at the market for law clerks, because it shows the relationship—and the difference—between moving too soon and moving too fast. After that market temporarily solved the problem of unraveling, by discouraging judges from making early offers, those judges found another way to use time strategically, and that led to the market unraveling all over again. This example makes it clear why, to succeed, a market has to solve multiple problems; it has to strengthen all the links in the chain of potential failures enough so that none of them will break.

Judges Delayed Are Judges Denied

> I received the offer via voicemail while I was in flight to my second interview. The judge actually left three messages. First, to make the offer. Second, to tell me that I should respond soon. Third, to rescind the offer. It was a 35 minute flight.
>
> —2005 APPLICANT FOR A FEDERAL JUDICIAL CLERKSHIP

This quote is from a law student who applied for a job as a law clerk to a federal appeals court judge. The calls he describes came while he was rushing from his first interview, in Boston, to his second one, the same day, in New York City. When he boarded a plane and had to turn off his cell phone, he still didn't have an offer. When the flight touched down thirty-five minutes later, he *no longer* had one.

When my colleagues and I studied this market, we promised anonymity to everyone we contacted, so some names of famous judges will be omitted here. My coinvestigators are not anonymous: Professors Chris Avery and Christine Jolls and Judge Richard Posner of the Seventh Circuit Court of Appeals in Chicago. Judge Posner is a longtime participant in the market in which federal appellate judges hire new law school graduates as their clerks. That market has undergone many changes in rules over the years, each of them an attempt to halt the kind of unraveling discussed in the previous chapter.

In each case, the judges tried to make a rule by setting a time before which no offer should be made. And in each case, the attempt partly succeeded, until more and more judges began to skirt the rules or ignore them entirely, and the market collapsed back into a race to hire first. As I write this, the most recent collapse has just begun, and by the time you finish this chapter, you will have a good idea of how it will likely end.

Over the years, as the judges have wrestled with unraveling, they've focused only on the problem of offers being made too early. And while they are eloquent about the problems this causes, they've viewed it as a simple matter of self-control. Many times they've formulated rules meant to control the date before which judges should not make offers to potential clerks.

But they haven't tried to control *how* those offers are made. To put it another way, by controlling the time at which offers are made, they make the market thick, but they haven't given the participants any tools to deal with congestion. Consequently, exploding offers that are open only for a shockingly short time have remained very common.

Judges aren't easily discouraged: they have proposed and tried similar rules about the timing of offers in 1983, 1986, 1989, 1990, 1993, and 2002, with the most recent attempt being formally abandoned only in 2013. After each attempt, the market unraveled, and before it was officially abandoned, more and more judges were ignoring the rules.

Yet the quote about exploding offers at the beginning of this section is from 2005, a year in which the market had *not yet* unraveled again: all offers that year were made around the same time. And that brings us face-to-face with congestion – insufficient time to make and consider as many opportunities as needed to make a good decision. Congestion is one motivation for unraveling.

Congestion is a little like rush-hour traffic in a big city. If everyone heads to work at the same time, a big traffic jam results, and it takes everyone too long to get to work. One way for an individual to deal with traffic is to head to work earlier. But if many people do that, the rush "hour" starts earlier and lasts longer. So everyone still spends too much time commuting, and some people may be tempted to start their commute even earlier. There are various ways to deal with traffic congestion collectively, such as better roads and bridges, which might let traffic move faster, or good mass transit, which might take some of that traffic off the road.

In virtual traffic jams like the ones that perennially plague the market for law clerks, exploding offers let the "offer traffic" move a little faster. But there are also less respectable reasons to make early exploding offers than just trying to make the offers flow. If a judge thinks that the person he is trying to hire might get more attractive offers if given more time, he can try to "capture" that person by insisting on an answer right away.

In 2005, the hiring of law clerks was supposed to focus on third-year students – that is, those who would complete law school and graduate at the end of the year. And it wasn't supposed to start until after Labor Day.

After many years in which law students had been hired before they returned for their second year of law school, the Labor Day rule went into effect in 2003. Thus by 2005, judges already had experience with this system. And what they had learned was that after Labor Day, their fellow judges—or at least some of them—would hire clerks *fast.*

So the judge making the calls in the earlier quotation knew that if he left his offer open very long and was eventually turned down, the next student to whom he wanted to make an offer would very likely already be gone. It wasn't *safe* for him to wait. Hence he made his offers fast and couldn't wait even thirty-five minutes for a response. (He may also have inferred that a student who didn't pick up his phone immediately might already be in another interview and was likely a lost cause.)

In a 1991 article titled "Confessions of a Bad Apple," Judge Alex Kozinski of the Ninth Circuit Court of Appeals defended what some other judges saw as his bad behavior in hiring clerks by saying: "This is the market that determines the career paths of some of the country's smartest and most promising young lawyers; it would be astounding if it were conducted with the gentility of a minuet. We are, after all, training courtroom gladiators not ballroom dancers."

Ballroom Dancers Meet Courtroom Gladiators

But even Judge Kozinski, who proudly jumped the gun, readily acknowledged the advantages to all parties—himself included—if hiring could be delayed. And yet despite six attempts to adjust the rules since 1983, each attempt has been followed by cheating and then wholesale defection from the rules. Following a rule change, judges might cheat only a little at first. But all it takes is a little cheating to open the floodgates. How do we know? The cheaters told us when we asked them in confidential interviews and surveys.

By 2004, one year after a new crackdown was implemented, fully 46 percent of judges responded in our survey that they knew of a substantial number of judges who hadn't adhered to the rules. By 2005, that number had gone up to 58 percent. Students surveyed about their own experience reported in 2004 that they had interviews scheduled and conducted before the allowed date. Although in 2004 only 12 percent reported getting an explicit offer before that date, the percentage had more than doubled by 2006.

As pressure on the rules mounted, many of the country's top law schools tried to intervene by telling their professors not to provide recommendations for prospective clerks before the agreed-upon date. But this was a hard rule for professors to follow, because their students would suffer if they refused to give a judge a recommendation when he called. A professor who said, "Sorry, Judge, I have a student who could be your best clerk ever, but I can't tell you about her until after Labor Day" would simply be telling the judge to hire the students of a more cooperative professor, perhaps from a competing law school. (Keep in mind that high-profile clerkships are a boon not only to the clerks themselves but to their professors and law schools, since such hires burnish a school's reputation and make it easier to attract top students in the future.)

This recruiting dance, and its relationship with the written rules, was for many law students their first intimate contact with judges in action, and for some of them it came as both a shock and a disappointment (maybe the kind of shock that "ballroom dancers" feel when they dance with "courtroom gladiators"). Their idealism about the integrity of judges was dashed. Some of the comments we heard from students reflected this:

> *It's sad (pathetic?) that judges aren't obeying their own rules. [It] flies in the face of the whole notion of "law and order."*

> *One of [Judge Z's] clerks even chastised me for "overly stringent adherence to this timeline they have" and noted that other students from my school were willing to interview ahead of schedule. It was a real conflict for me. I felt like I had to choose between cheating and (potentially) not getting a clerkship.*

> *It's very disheartening to see so many Federal judges — the ostensible paragons of rules and fair play — breaking their own rules . . . I expected better.*

Note that judges who chose not to follow the rules went only a *little* early. Cheating "just a little" was in fact the most attractive option, since it meant that a judge could make his offers and hire his clerks before his competitors, but without sacrificing the important information about candidates that he would do without if he went substantially earlier.

It's also worth noting that the most enthusiastic cheaters weren't the most prestigious judges in the most prestigious circuit courts, such as the D.C. Circuit. But neither were they very far down in the pecking order. Rather, the judges who first moved up their hiring dates, and who from year to year nudged them just a little earlier, tended to be the top judges in slightly less prestigious courts — such as the California judges in the Ninth Circuit, like Judge Kozinski.

It's not hard to understand why this should be the case, since the judges at the very top of the pecking order are the ones who will always benefit the most if hiring happens only after all information about candidates' law school careers is available. That's because if all judges wait, the best judges get to hire the best candidates — so why should they cheat?

On the other end of the scale, it wouldn't work for a low-prestige judge to make early offers, since the top candidates would turn him down if taking a clerkship with him meant forgoing the chance at a better offer. So those judges, even if they tried to hire earlier than everyone else, probably wouldn't have much success.

But the *almost*-top judges faced a very different set of choices. If they waited to compete with the very top judges after all the information about students was in, they wouldn't get to hire the top law review editors and other prizewinning students. Although the almost-tops could still expect to hire very good clerks, their chances of hiring clerks who might subsequently be hired by one of the nine Supreme Court justices would take a big hit.

It takes a very confident (or foolish) student about to begin her third year of law school to turn down a plum job just because it isn't the plummiest job in the whole country. By hiring a top student at the beginning of her third year, an intrepid almost-top judge has a chance of hiring someone who might turn out later to be one of the big winners—which might even get that "almost" removed from the judge's reputation.

So it was the almost-top judges who started the unraveling. And when they were successful in getting too many of the top students, the very top judges had to hire early, too, in self-defense.

Thus year after year, there was more and more cheating, until some of the law schools quietly stopped telling their professors not to comply with early requests for information about clerkship candidates. Stanford University, home of a top law school, was late but very public when it wrote to the Judicial Conference of the United States in June 2012 to say that in light of the widespread violation of the rules, it would henceforth release letters of recommendation whenever they were requested.

The Stanford letter indicated just how far the cheating had progressed. It said: "Increasing numbers of judges—the entire membership of some courts, some or many of the judges in most others—have begun interviewing and hiring law clerks well before the Plan."

The official end of the hiring rules—"the plan," as they were called collectively—came in January 2014, when even the most prestigious circuit of all, the D.C. Circuit, ruled that the scheme was history by publishing a notice that said in part:

> Although the judges of this circuit would uniformly prefer to continue hiring law clerks pursuant to the Federal Law Clerk Hiring Plan, it has become apparent that the plan is no longer working . . . We stand ready to work with the judges of the other circuits to develop an appropriate successor to the current plan.

> In the meantime, however, the judges of this circuit will hire law clerks at such times as each individual judge determines to be appropriate. We have agreed that none of us will give "exploding offers," that is, offers that expire if not accepted immediately.

Those few sentences say a lot about prestige and its limits. The D.C. Circuit, the most prestigious in the country, had the most to gain from hiring students as late as possible. It was the last to abandon the rules intended to promote later hiring, and it was eager to restore those rules in the future—not just out of fairness, but because it stands to gain the most from an orderly market. While the D.C. Circuit appeared to be making a noble sacrifice by agreeing not to make exploding offers, judges in the most prestigious circuit don't need to make exploding offers. Unlike judges on other courts, they are confident that few students will turn them down. But having the most desirable positions to offer didn't protect them from having to hire earlier than they would have liked once everyone else was doing it.

Thus judges on the D.C. Circuit promptly moved the hiring of 2014 clerks before Labor Day. Judge Janice Rogers Brown, for example, was widely reported to have hired a clerk named Shon Hopwood in the first week of August 2013. Hopwood has an unusual personal history: before entering law school he served a lengthy prison sentence. But his early hiring quickly became quite usual. Clerks who wouldn't begin work until 2015 were being hired in February 2014, a year and a half early.

I predict that the unraveling of the law clerk market will continue and that in the coming years, judges on the D.C. Circuit and others will once again find themselves hiring clerks well before the middle of students' second year of law school—and law students will find themselves facing exploding offers.

The history of this market is a case study in how incremental changes in design can fail when they address only the *symptoms* of market failure and not the *causes*. In this case, the change was a set of rules that said that judges should wait to make their offers — even as the actual market didn't make it safe for them to do so.

Slow-Motion Explosions

As we've seen, judges wield exploding offers as easily as they wield their gavels. Exploding offers happen in many markets, but sometimes they take forms that might be hard to recognize. In one Japanese market, these explosions took place over the course of months.

The essence of an exploding offer is not that it expires quickly, but that it forces the recipient to respond before receiving any other offers. As far back as 1953, to stop unraveling in the hiring of Japanese university graduates, attempts were made to specify a time before which job offers should not be made to students. That year, universities, business organizations, and government ministries agreed that schools shouldn't begin to recommend seniors to companies until October 1. (The Japanese academic year begins in April and ends in March, so this meant the second half of their senior year.)

As in so many cases, mandating dates didn't work, but those who had a stake in the market didn't stop trying. The 1970s were marked by a series of agreements between firms, university organizations, and government ministries concerning dates at which various recruiting activities could be undertaken.

These agreements failed for reasons we can now readily guess at: some potential employers simply ignored them or found clever ways to circumvent the rules, such as offering early informal guarantees of employment. As some firms made early offers, other firms felt pressed to make even earlier ones. In the 1980s, the Labor Ministry announced that it would no longer monitor the current agreement, since it had no effective way of enforcing it. (Recall how the NCAA gave up trying to regulate the timing of agreements for football bowls, discussed in chapter 4.) But amid all this rule breaking,

Japanese companies nevertheless found it embarrassing to issue *official* offers of employment before the specified date.

Instead, they made a kind of slow-motion exploding offer, by essentially kidnapping the students to whom they had made early informal offers. For example, they scheduled mandatory events on the days when civil service exams were held. If a student didn't come to the company event, he wouldn't be given the promised offer when the day for official offers came. As a result, a student who wanted to take an early offer couldn't take the exam that would allow him to get an offer from, say, the Finance Ministry. In essence, the student was holding an exploding offer that he would have to turn down if he wanted to consider other options.

A 1984 survey showed that 88.4 percent of Japan's major companies thought that the current agreement on recruiting graduates should be continued, although 87.7 percent admitted that they did not abide by it.

In the Japanese market for new graduates, as in the American markets for new lawyers seeking positions in law firms or as clerks, it was hard to create a thick market in which many people could have multiple opportunities from which to choose. It wasn't sufficient to adopt rules that stopped offers from being made early. Even when offers were all made around the same time, they turned into exploding offers, which still didn't allow those receiving them to consider multiple opportunities. Before participants can enjoy the benefits of a thick market, a marketplace has to overcome the congestion that thick markets bring with them — that is, the problem of how to allow multiple offers to be made and considered in the time available. We turn next to consider this problem of congestion.

6

Congestion: Why Thicker Needs to Be Quicker

WHEN IT COMES to speed, markets tend to follow a kind of "Goldilocks principle": they mustn't be too hot or too cold.

We've just seen how too-fast transactions can hurt a market. But excessively slow transactions can hurt markets, too. Surprisingly, markets can be too slow, or *congested,* even on the Internet. Although the Net operates at the speed of computers, the people using it still need time to consider and act. That's why, if you really want to operate at digital speeds, you need to take people out of the middle of the process. One way to do this is by moving their deliberations to an earlier time. (Hence the emerging Internet of Things, in which devices learn your preferences, talk with one another, and make decisions for you.)

Markets that involve offers and responses require easy two-way communication. This is why the rise of mobile communications has been so important for the development of many Internet markets: smartphones shorten response times.

Consider Airbnb, which makes a market between travelers looking for a nice, cheap place to stay and hosts who want to rent out their underused guest rooms and apartments.

When Airbnb started in San Francisco in 2008, most people communicated with the Internet via computers. So if you wanted to make your guest room available for visitors the following week, you might use your laptop to post it on Airbnb in the morning before leaving for work. When you came home in the evening, you would check to see whether anyone had expressed an interest—and if so, you would confirm his or her booking. Easily done.

Now look at the other side of this transaction. As a potential guest, you might have had to wait a whole day to find out whether the room you wanted was still available, and if you learned at the end of the day that it wasn't, you would have had to start over.

You can see the problem. Airbnb's business model worked well enough in the beginning, when the market was small and the travelers were intrepid young people on tight budgets who were willing to take the time to find a good deal. Airbnb's competitors in those days were similar Web services, such as the London-based Crashpadder (acquired by Airbnb in 2012) and the Toronto-based iStopOver (acquired by the Berlin-based 9Flats, also in 2012). Competition in those days was based largely on attracting more and more hosts and travelers in order to make the market thicker. That's why bigger fish acquired littler fish.

But as Airbnb became huge, with lots of hosts and travelers, it became increasingly common to have to make multiple attempts to nail down a reservation. Meanwhile, Airbnb's main competitors were no longer other small Internet businesses, but giant hotel corporations such as Hilton, Marriott, and Best Western. And one huge advantage these huge hotel chains offer to travelers is speedy confirmation. Their transactions are fast: by phone or on the Web, you can quickly find out whether rooms are still available and book one for the night you want. That's because all the rooms in, say, a Hilton are managed by a central computer system, so one call lets you check all the rooms at the same time.

Imagine instead if you had to call Hilton to inquire about each room individually. On any given call, the only thing the reservation clerk could tell you was whether, say, room 1226 at the San Francisco Hilton was available for the night you wanted. If not, you had to make another call to find out about room 1227, then another for room 1228. Booking a room with an Airbnb host was a little like that.

So Airbnb had to figure out how a market with many hosts offering one room at a time could compete more effectively with hotels. Price was obviously important. But it was the spread of smartphones that helped Airbnb close the speed gap, and that may have mattered even more than price. Today, as hosts manage their reservations on their smartphones, they don't have to wait until they return home to confirm a booking—they just check their phones. They can also, as soon as the room is booked, immediately update their Airbnb listing to remove its availability. That in turn makes it easier for a traveler searching for a room to find one that's available, even though he or she still has to query one room at a time.

Thus smartphones make the home hosting market work better not just because hosts can respond faster but also because they can update their bookings, which makes them more informative. This, too, reduces congestion (fewer rooms appear to be available, and a room that looks available is more likely to actually be so), and as a result helps travelers search more efficiently, with fewer time-wasting false leads.

Just as Airbnb competes with hotels by making a market for bedrooms, the transportation company Uber competes with taxis by making a market for limos and private cars. In most cities, only taxis have the right to pick passengers up on the street, while limousine services can pick customers up only if they have made a reservation in advance. Those reservations used to take time to arrange. So although limos worked fine for a scheduled trip to the airport or for a fleet of black cars for a conference, if you stepped outside and it was raining, or checked out of a hotel after a leisurely breakfast, hailing a taxi was much easier.

Once again, smartphones changed all that. Now you can call a limo almost as easily as a taxi. So a lot of limos that once sat idle are now readily available. And that's just the beginning. UberX and companies such as Lyft are even more like Airbnb: they are starting to make a market for the vacant passenger seats in private cars. Speed is of the essence in making these markets work differently than the "day ahead" market that already existed for limos. In some respects, these services are faster even than taxis: you don't have to take time to pay when you reach your destination, since the same app through which you called the car will also pay the bill automatically through your credit card. This works because your original call goes through a central clearinghouse that archives your credit card details.

Perhaps you are beginning to see a pattern here and can anticipate (even build) marketplaces that don't yet exist. Limousines have been around for a long time, and there have always been spare bedrooms that you could arrange to rent through friends. But computers and smartphones have helped Uber and Airbnb build multibillion-dollar businesses by making those markets thicker and quicker, bigger and less congested.

Opportunity Knocks

The opportunity to do well by building a good marketplace can arise whenever there are desirable but underused resources that take too much time to find and transfer.

How about the tickets you bought for a game or a show that you can no longer attend—or tickets that you really want but are sold out? StubHub is now making a market for those tickets. In 2007, StubHub was acquired by eBay, which may be the first Web-based company to make a market for things that were sitting idle in someone's garage or attic. (By the way, speed is becoming important to eBay, too. Whereas most items were originally sold by auction, today most are sold at a fixed price. That's a faster way to do business, because you can buy what you want as soon as you want it, without having to wait for an auction to end—and taking the chance of losing and having to try again in another auction.)

Reservations at fancy restaurants, which are often booked far in advance, are similar to unwanted tickets. No companies have yet made a large-scale market for restaurant reservations as I write this in mid-2014—you're still pretty much stuck with scalpers and concierges—but some start-ups are giving it a try, in cooperation with the restaurants whose reservations they'll offer. Stay tuned.

And just as your spare bedroom might have languished unused before Airbnb made a market for it, your home Wi-Fi might be lying idle during the day while you are at work. If I'm your guest, I can ask you for your password. But if I'm driving by your house, even though my phone may detect your protected Wi-Fi, I can't access it, much less pay you for its use. Yet as I write this, a start-up called BandwidthX (I'm on its advisory board) is figuring out how your cell phone provider might be able to access this "Dark Wi-Fi" automatically. This could get travelers better connections when they have a weak signal or the system is overloaded, and it could get home owners reduced rates on their Wi-Fi by letting them sell access to it.

In all of these new markets, the entrepreneurs building them have had to figure out

1. how to make the market thick by attracting lots of buyers and sellers;
2. how to overcome the potential congestion that could result—that is, how to make the market quick even when it was thick; and
3. how to make the market safe and trustworthy (we'll return to this later).

For Uber and Airbnb, dealing with congestion was a make-or-break issue. If making a match between cars and passengers became too frustrating, lots of passengers would go back to using taxis. If it became too hard to book a room in someone's house, travelers would go back to booking hotel rooms. Congestion threatened the very thickness of the market that makes them both big businesses.

It can be hard to notice how dangerous congestion is to a market, because most successful markets have found a way to deal with it, and most markets that can't deal with it fail to become big and thick enough for us to notice them. But congestion makes markets work badly even when it doesn't threaten their ability to remain thick, with an abundance of participants on both sides.

Consider the matching market for putting children in public schools. New York City has lots of schools and lots of students, and the New York City Department of Education doesn't have to worry too much about losing either. But the DOE did have to find a way to deal with congestion before it could do a good job of deciding who gets assigned to which schools. By looking at New York, we can see clearly the problems that congestion can cause and how they made summers hotter and longer for students waiting to be assigned to schools.

If You Can Make It There . . .

If markets can be too slow even using the Internet and smartphones, imagine how congested they can be when conducted by mail. In comparison to newer forms of communication, the mail has a leisurely quality to it. Even if people wait impatiently for an important letter to arrive, they may still feel that they can take a little time before replying.

I learned of just such a problem when my phone rang one day in 2003. On the line was Jeremy Lack, director of strategic planning for the New York City Department of Education. Jeremy's boss, Chancellor Joel Klein, had been charged by Mayor Michael Bloomberg with reasserting mayoral control over the largest municipal school system in the country.

New York's public schools had operated in a decentralized way for years, with autonomy given to individual school principals and community school boards. Mayor Bloomberg and Chancellor Klein aimed to give the almost 90,000 new ninth graders each year a meaningful choice of which of the city's hundreds of high schools they might attend.

Shortly after that, Parag Pathak, a star graduate student who'd taken my course on market design at Harvard, came to my office in search of a project. He told me that he wanted to combine economic theory with "something real." I suggested that he help me dig into the New York schools. Today he's a professor of economics at MIT and an expert on school choice. We also drew in Atila Abdulkadiroğlu, who's now a professor at Duke University but was then conveniently located in New York at Columbia University.

With so many students and hundreds of schools, the market for school places in New York was plenty thick. But it was also unbelievably congested.

In those days, New York ran a complicated paper-based high school admissions system. Children preparing to enter high school, assisted by their families, filled out forms listing up to five high schools they liked in rank order. The DOE collected these forms and sent copies of them to each of the schools listed. These became the students' applications to the schools.

Some schools were required to admit students by lottery, but others could choose whom to admit or wait-list. After the schools had decided, the DOE sent letters informing the students where they'd been admitted. Each letter asked the applicant to choose a school (if the student was admitted to more than one) and a waiting list (if the student wished to remain on such a list for a school he preferred over the one to which he'd been admitted).

The rules also stated that students could accept no more than one offer and one waiting list. Schools that had some first-round offers rejected could then make new offers, and the DOE sent out a second round of letters. After students replied, there was a third and final round of letters and replies. Students without seats after the third round were assigned by the DOE, typically to the school with open seats closest to their home.

The result of this complicated system was a mess. Many students couldn't get seats in any of the schools to which they'd applied—and when they were assigned, it was usually in August, just before school began. Many students didn't take part in the process at all but instead, amid the chaos and congestion, slipped into schools through unofficial channels.

The reality was that three rounds of processing applications for 90,000 students just couldn't place everyone.

Meanwhile, the system was also congested: there wasn't time for more than three rounds. Only about 50,000 students received offers in the first round, and of them about 17,000 received multiple offers that had to be accepted or rejected before the next round could start.

You can see how that would slow things down. Even if you were accepted to your first choice, sending in your letter of acceptance in a hurry might be a low priority for you. Instead, you might celebrate for a day (or three) before mailing your reply to the DOE. And if you weren't accepted to your first choice, you might want to consult with neighbors, friends, teachers, and others before deciding which waiting list you should stay on—and that could take time, too. So the problem wasn't just that "snail mail" can be slow but also that decisions take time. And a lot of decisions had to be made by a lot of people as the process played out.

By the time the third round concluded, about 30,000 students still hadn't been accepted to a school on their list. So they had to be assigned somewhere by the central office. Think about that: 30,000 kids and their parents, sweating out one of the most important events to date in their lives, right down to the last hot moments of the summer before entering high school.

That was frustrating enough, but congestion wasn't the only problem. To many parents, the whole system seemed risky, unsafe, and untrustworthy. There was, officially, an appeals process and an "over-the-counter" process for assigning students who had moved or were otherwise unassigned before school began. But savvy parents knew that they could also appeal directly to principals, since schools didn't have to include all their available places in the centralized process. The result was a robust "gray market" of motivated parents circumventing a system they viewed as opaque at best, and biased or corrupt at worst.

"Joel sensed that kids were getting in schools based on things other than the merits of their applications," says Tony Shorris, deputy chancellor for operations. "Well-connected kids had an advantage."

Another aspect of the old system that made choosing schools tricky was that the principals saw the students' preference lists, so they knew where the students had ranked their schools. In response, a lot of schools would not admit any student who had not ranked them first.

Given the specialized nature of some of New York City's high schools, this might seem sensible. Say you were the principal of Aviation High School, located between New York's LaGuardia and Kennedy airports. Your mission was to prepare students for careers in aviation. You might have wanted to accept only those kids who felt called to that industry, and you might have assumed that they'd indicate this interest by ranking Aviation first. But if you restricted admission to those students, no one who ranked Aviation second could get in, and some students for whom Aviation was in fact their second or lower choice would face the strategic decision of whether to rank the school first nevertheless.

A student whose genuine top school was hard to get into would give up any chance of going to Aviation if she revealed her true preferences. The DOE implicitly acknowledged this in its 2002–3 directory of high schools, which advised that students, when ranking schools, should "determine what your competition is for a seat in this program." In other words, the school system was telling kids and their parents to calculate and strategize, not just think about which schools they liked. This meant that even if Aviation's principal admitted only students who ranked the school first, he had no guarantee that it really was their first choice. In truth, *it was only the best school they thought they could get into if they ranked it first.*

Families weren't the only folks strategizing. The system wasn't safe for school principals either, who often felt pressed to game it by concealing capacity, and holding back empty seats, until after students had been assigned. They then tried to fill those seats with students who weren't happy with their assignments. The *New York Times* later quoted a deputy chancellor as saying, "Before you might have had a situation where a school was going to take 100 new children for 9th grade, they might have declared only 40 seats and then placed the other 60 children outside the process."

But the most pressing problem, the one that prompted the call to me for help, was the massive crowd of 30,000 students who could not go to *any* school they'd chosen and instead had to be assigned, administratively, at the last minute. And that, too, was a problem of congestion, of not enough time to make enough offers of admission, and enough acceptances and rejections, to reach every kid who needed a spot, while still allowing families to have input on their kids' fate.

Decongestant

Thick markets need to be quick, but it's hard to be quick—no matter how fast the technology—if people have to wait for other people to make and act on their decisions. At the beginning of the New York City high school choice process, schools could make offers without waiting for anyone. But once they made some offers, they had to wait for replies before they could make new offers, and that caused the market to be congested, and too slow to process every student in time.

Consider how this works in the market for houses. When you're *selling* a house, the asking price is offered to everyone. And when you're *buying* a house, you can consider every house on the market.

But suppose you decide to make an offer on a particular house. A typical offer comes with "earnest money" and is a signed offer (backed by a deposit) to buy the house at a specified price. To give the seller time to consider the offer, it typically also comes with a specified duration, perhaps twenty-four hours—a little longer in a slow market and shorter in a hot market like the one in which I now find myself in Silicon Valley.

During the time the offer is open, most buyers can't afford to make an offer on another house, so they must wait for a response. But in a hot market, while you're waiting, another house on which you might want to make an offer may get sold. Similarly, if the seller makes a formal, signed counteroffer, he will have to wait for your decision, and he doesn't want to leave it open any longer than is necessary. So although the market starts with everyone talking to everyone else, it abruptly turns into a series of private conversations between one buyer and one seller.

We've seen how job markets can be like that, too. A company looking to fill an important position can survey the market and interview a lot of candidates, then consider to whom to make an offer. But once the offer is made, the company often has to give the candidate a little time to consider it, and during that time the other candidates might take jobs elsewhere and become unavailable. As we saw in chapter 5, that's especially true in a market in which offers are exploding and everything happens faster, so that companies may feel compelled to make exploding offers themselves.

Matching markets often have to deal with congestion, since each offer isn't just a set of terms, it's a proposal of a match to a particular counterpart. (In the market for houses, a buyer may not care deeply whom he buys from, but he certainly cares which house he buys, so markets for houses need to make matches, too.) Commodity markets usually have somewhat less trouble with congestion, since the offer to buy or sell shares of stock or bushels of wheat can be made to the whole market, and the buyer or seller can change a bid or an ask at any time, without having to wait for anyone else. But in matching markets, some offers have to wait for the decisions of others.

The key to speeding up the New York school choice system wasn't just to design a computerized process, although that was part of it: computers are fast. But people are slow, and both the decisions that people make and the information they have about which schools would be best for their kids are critical to the market's success.

The solution to decongesting New York's high schools turned out to be letting people indicate their school preferences all at one time and then using those preferences to process decisions quickly. When preferences are submitted in advance, a computerized clearinghouse already knows which school a student would prefer if he was given a choice of two.

The same system can work in other matching markets, such as job markets, in which an applicant can indicate her preferences among the jobs she has applied for all at one time.

With various colleagues, I have helped design computerized clearinghouses that are now used to process preferences for schools, jobs, and workers that people decide on *before* the computer is turned on. The computer can then run through offer-rejection-offer chains fast enough that there's time for every offer that someone wants to make to be made.

The key to such a clearinghouse is making it *safe* for people to state their preferences honestly. So before I start to describe the clearinghouses we've built, let's talk more about safety.

7

......

Too Risky: Trust, Safety, and Simplicity

MAKING MARKETS SAFE is one of the oldest problems of market design, going back to well before the invention of agriculture, when hunters traded the ax heads and arrowheads that archaeologists today find thousands of miles from where they were made.

More recently, one of the responsibilities of kings in medieval Europe was to provide safe passage to and from markets and fairs. For healthy commerce, buyers and sellers needed to be able to participate in these markets safely, without being waylaid and robbed (or worse) by highwaymen. Indeed, the word *waylay* captures the act of robbing travelers carrying money or valuables on their way to or from a market. Without some assurance of safe passage, these markets would have failed; they would have been too risky to attract many participants. And if the markets had failed, the kingdoms would have been deprived of the prosperity that markets, and the taxes on them, bring.

In the chaos of the Oklahoma Land Rush, discussed in chapter 4, even those Sooners who managed to claim desirable land found the trip to register their claims at the land office in Enid a dangerous one. Many claimants chose to ride together in large groups to discourage ambushes along the way.

This risk of robbery and physical harm still exists today in some markets, particularly in illicit ones, such as those for illegal drugs and sex, where buyers and sellers often meet furtively in isolated, poorly policed places. The risk comes not just from third parties who might prey on the markets and steal money or goods; it also comes from sellers and buyers who sometimes prey on one another. Indeed, one of the long-standing arguments for legalizing drugs and prostitution is that outlawing these markets merely moves them into the unregulated and unsafe criminal world. (I'll have more to say about this in chapter 11.)

Of course, it's not only illegal markets that can be dangerous. For example, taxi drivers in big cities face some risk from passengers who ask to be taken to poorly policed neighborhoods with the aim of robbing the drivers. And you have only to watch the news to know that working in retail—jewelry stores, banks, gas stations, and convenience stores—can make people vulnerable to robbery. Just as medieval Europeans did, modern-day citizens count on governments to provide the kind of basic security that markets can't thrive without.

But there are other, more prosaic risks that come with markets. You might not get the goods you paid for, or they may not be of the quality that you expected. Or your credit card information might be stolen and used to purchase things for which you will be billed. (It is precisely to make it safe to use a credit card that the issuing bank indemnifies you against this kind of financial loss—though not against the inconvenience of dealing with fraudulent charges.)

This is why buying something in a legal marketplace from an easily identifiable merchant is almost always safer than an illegal transaction. A legal market gives you some confidence that you'll be treated fairly, especially because if you aren't, you have potential recourse to the law.

Years ago, when my wife and I lived in tiny Farmer City, Illinois (population 2,000), we put down a deposit on a dining room table at a store located in a strip mall in a bigger town. Months later, after many fruitless attempts to get either the table delivered or our deposit returned, I made my way to small-claims court in the county seat. The clerk gave me a form to fill out and helpfully explained that I could copy the store's address from one of the many other complaints that had been filed against it.

After several trips to court, due to delaying tactics by the furniture store, I eventually got a judgment in my favor. We even got our deposit back—just before the store went out of business. So my encounter with this dishonest or disorganized seller was merely annoying, and I may even have played a small part in improving the local marketplace by helping him leave the business. (Of course, he may have just moved to another town and started taking deposits again, until he earned a bad reputation there, too, and had to move once more.)

Reputations are important. If I returned to that corner of Illinois today, I would probably find that some of the furniture stores there now have been in business since before I left three decades ago and are on their second or third generation of customers. Their longevity would provide at least some testament to their honesty and reliability. Any store that's been in one place for a long time probably has a reasonably good reputation, and if you have doubts, you can pretty easily ask around to find satisfied or dissatisfied customers.

When everyone is in the same place, good reputations can be earned naturally by honest businesspeople. But (to paraphrase a famous cartoon) on the Internet no one knows you're a reputable businessman. As a customer, it's a whole lot harder to assess the reputation of someone whom you know only by a username and whose other customers you've likely never met. And if you're setting up shop online, you have to find a way to convince prospective customers that you're trustworthy and not some con artist operating out of an Internet café on the other side of the world.

That's why, until a few years ago, one of the biggest problems facing new Internet marketplaces was how to make it safe to transact with strangers—and how to convince customers of this safety.

For example, the challenge for eBay—particularly after some well-publicized scams—was how to make millions of customers each day feel confident that they would get what they ordered. In much the same way, if you were a host renting your room to a stranger on Airbnb, a driver answering a call for a car on Uber, or someone selling some item to a stranger on Craigslist, you would be glad to have some assurances about the buyer. And if you were the buyer, you might want some assurances about the host, the driver, or the private seller.

So far, I've mostly talked about market safety. But I also want to put in a word for *reliability.* Both safety and reliability fit under the general heading of making a market *trustworthy.* When you order a car on Uber, you want to know not just that you will get a safe driver and that the car won't be a wreck, but also that the car will arrive promptly. Just as important, before you download the Uber app on your smartphone, you want to know that the system won't be buggy, slow, or inaccurate (that is, the car will be able to find you). And you want to be able to provide your personal information, including your credit card number, without worrying about identity theft. If the Uber app had fallen down in any of these areas, customers would have quickly purged it from their phones, and the company would never have survived.

In much the same way, the Uber driver wants to know that *you* are reliable—that you won't call a taxi without canceling your call to him and leave him looking for you, and that his charges will be properly paid after you leave the car.

So for a marketplace to be truly trustworthy, it must be safe; participants on both sides of a transaction must be able to rely on each other and on the technology.

A Good Name

Approaches to making Internet markets trustworthy and secure are constantly evolving (in part because the bad guys are clever and adaptive themselves). Up to this point, market designs for trustworthiness have focused on providing secure methods to make payments, providing insurance for transactions that go bad, and building feedback systems that allow reliable sellers, and sometimes buyers, to develop and display good reputations.

eBay was one of the first Internet markets to tackle these issues. In the early days, many sellers didn't have an established, physical business that might give them a reputation they could carry over to the Internet. Almost everyone had to start building a reputation from scratch.

The issue of trustworthiness affected not just sellers (who might not provide the goods as described) but also buyers (who might not pay promptly or at all). It didn't take many bad checks for sellers to demand cash or money orders, or to wait until a check cleared the bank before shipping.

These concerns motivated the original eBay feedback system, which was set up before the introduction of convenient online payment mechanisms such as PayPal. This system was designed to allow both sides of a transaction, the buyer and seller, to provide feedback about each other that would be available to future potential transactors. The initial feedback rules, which involved leaving a positive, neutral, or negative rating as well as a text comment, soon underwent some modifications based on experience. The new goal became to stop people from artificially inflating their positive feedback. The system eventually evolved into one in which feedback was identified by the username of the person leaving it, and only the winning bidder and the seller could leave feedback about each other. That way, ratings couldn't be easily distorted by "feedback stuffing" from one individual.

Even so, over time the ratings became overwhelmingly positive, which made them less informative. Why this happened shows us yet again how careful attention to the subtleties of markets can highlight aspects of human behavior that might otherwise remain hidden. In this case, most of the feedback had become *reciprocal*, with sellers leaving feedback for buyers that mirrored the feedback the buyers left for them. Buyers and sellers were adhering to an unwritten rule of eBay culture: you scratch my back, and I'll scratch yours. The result was that the vast majority of the feedback following a transaction was mutually positive, with just a smattering of mutually negative posts.

With the help of three economists — Gary Bolton, Ben Greiner, and Axel Ockenfels (all former students or postdocs of mine) — eBay designed a new feedback system in which buyers could anonymously give sellers more detailed feedback on how accurately an item was described and how quickly it was shipped. This system improved the informativeness of the ratings. Suddenly it appeared — surprise! — that not everyone was thrilled with every transaction.

Here eBay's experience revealed yet another principle of good market design. Markets depend on reliable information. In the case of reputation, buyers want reliable information about a seller, in the form of information about other buyers' experiences. But if it is costly or risky for buyers to supply that information, they won't do so, and the whole market will suffer. When feedback on eBay wasn't anonymous, a dissatisfied buyer who did other buyers a service by making his dissatisfaction known ran the risk of retaliation from the seller. Particular buyers and sellers may have been benefiting (or, in some cases, punishing) each other by giving reciprocal feedback, but they hurt the market as a whole by limiting the supply of accurate information. When eBay made it safer to reveal dissatisfaction, the information about sellers became more detailed and useful.

When I visited eBay in 2014, I learned that the company was contemplating further changes to reflect the evolution of the marketplace. More and more sales are now coming from professional merchants selling new goods, rather than individuals selling used treasures found in their basements and attics. As the market transforms in this way, the insurance that eBay provides on individual transactions may relieve buyers from much of the risk they might otherwise face. Meanwhile, eBay's ability to monitor the performance of increasingly professional sellers may allow it to display the best performers more prominently, while removing dishonest or incompetent ones. That is, as professional sellers start to play a larger role in the market and become more like conventional stores, eBay may be able to more effectively protect buyers from bad sellers the way that shopping malls do—by evicting those that give the marketplace a bad name.

Too Much Information

While markets need a lot of information to work well, there can sometimes be too much information sharing, since market participants want and need to preserve some privacy. Failure to give them enough privacy can make the market unsafe, which in turn can make it fail.

The simplest example is the privacy of bank accounts and credit card numbers. Internet communications used to be even less private than they are now, but even today you need to be careful with some of your personal information. PayPal solved the payment privacy problem for eBay and other Internet markets by providing a payment mechanism that doesn't require you to type in your credit card information for each transaction. But you may be reluctant to reveal other kinds of information, too, and that can interfere with the operation of the market.

For example, in an eBay auction, you're invited to submit a proxy bid of the maximum amount you're willing to pay. eBay promises to use that number to bid on your behalf, but only as high as needed for you to be the winning bidder. Of course, you have to trust eBay to use the information as it promises (and not just charge you the highest price you're willing to pay however the bidding went). Since this kind of trust is essential to eBay's business model, I'm not surprised that I've never heard any suggestion that eBay has ever abused it.

But there are other reasons for reluctance about revealing your willingness to pay. Other bidders (or unscrupulous sellers pretending to be buyers) might test your bid by raising theirs to see how you (automatically) respond – in order either to find your ceiling or just to drive up the price. You may prefer to turn the tables by waiting, letting another bidder make the initial bid – and thus becoming the high bidder – and then swooping in at the last moment with a surprise higher bid, too near the scheduled end of the auction to provoke a bidding war (as an early bid might have done).

In fact, many eBay auctions are "sniped" by a bid placed in the very last seconds of the auction. There is even sniping software that can help you automate the process.

You might think of sniping as the opposite of unraveling, since it occurs very late, rather than too early. But both phenomena suggest that the main market has become risky in a way that makes it worth taking a different kind of risk. In eBay auctions, many bidders feel it's too risky to reveal early how much they're willing to pay. You can avoid that by taking a different risk – by planning to bid at the last second. The risk is that sometimes you forget to bid, or your bid comes in too late and isn't recorded.

Axel Ockenfels and I surveyed eBay snipers when eBay was new. We found that almost all of them had experienced both kinds of failures when they planned to bid in the last moments of an auction. Nevertheless, they felt that bidding at the last second was safer than running the risk of revealing too early how much they were willing to pay (hence the market for sniping software).

When participants in a market are reluctant to reveal crucial information, the market may run inefficiently. On eBay, concealing bid information from other bidders by sniping makes prices unpredictable, and when there's a lot of sniping, not every auction is won by the person who is willing to pay the most.

Sniping allows people to bid on eBay while protecting their information, without striking a lethal blow to eBay's marketplace. But other marketplaces have failed because they tried to force participants to reveal information they didn't feel safe broadcasting.

For example, Covisint was started in 2000 by a consortium of the biggest car companies. It was intended to be a transparent online marketplace for automobile makers and their suppliers. But it turned out that auto parts suppliers weren't wild about making their prices public to auto companies and competitors. By 2004, the auto makers threw in the towel and sold Covisint for a tiny fraction of what they'd invested in it.

A related problem befell a Pittsburgh-based auction company called FreeMarkets, which set out in 1995 to change the way companies bought supplies. It offered to run procurement auctions in which potential suppliers would bid by indicating the costs they would charge for particular orders, with the lowest bidder being the winner. FreeMarkets offered an additional service, which was to find new qualified bidders for its clients' needs. The idea was that if a company defined the things it needed to buy very precisely, as commodities, it could buy them all at auction from a larger set of potential suppliers, and thus drive its costs of procurement way down.

Things didn't work out the way FreeMarkets hoped, in large part because many companies aren't only buying commodities, in which price is the only important dimension. They are often in matching markets rather than commodity markets, because they are in long-term relationships with their suppliers. Those suppliers felt it was risky to reveal their customary discounts and business practices to competitors, and so procurement auction markets failed to catch on.

During the height of the Internet boom, FreeMarkets' stock

market capitalization was briefly greater than that of its Pittsburgh neighbor U.S. Steel. But that didn't last, and FreeMarkets was sold in 2004.

Marketplaces as varied as eBay, FreeMarkets, and the New York City public school system reveal a challenge that must be faced by virtually all markets: how to manage the flow of information. No matter how well a market is otherwise designed, it will have trouble giving people what they want if it doesn't make it safe for them *to try to get* what they want.

Boston Public Schools

In Boston, the system for assigning school places flouted this principle in a big way. But the system's biggest problem was that it didn't even realize it had a problem.

In contrast, redesigning the New York choice system was in some ways like treating a heart attack: the patient understood that something had to be done, and quickly. There was no option to delay —those 30,000 unmatched students couldn't be ignored. The big problem in New York schools was congestion. The fact that it wasn't safe for families to reveal their preferences seemed secondary.

By comparison, solving Boston's market design problem was more like treating a patient with high blood pressure. That's a dangerous disease, too, but its symptoms are more subtle.

Unlike New York, Boston already had a smoothly running computerized choice system, in which families rank-ordered lists of schools and children received admission to a single school. So congestion wasn't an issue; everyone was assigned quickly, although waiting lists for preferred schools moved slowly after the main assignment was finished.

Boston Public Schools (BPS) used its assignment algorithm not just for high schools but also for kindergartens and middle schools.

And it seemed to assign many children to their parents' first choice of schools. So far, so good.

But these positive results disguised another problem: *users didn't trust the system.* BPS tried very hard to give families what they wanted, but how it did so made it too risky for those families to reveal what they really wanted.

Underlying Boston's system were rules that determined the priority that each child would be awarded for admission to each school. For half the seats in a typical school, this priority went first to children who had older siblings attending the school. Then priority was given to children who lived in the school's "walk zone." A random lottery number was assigned to each child and used to break ties —for example, when there weren't enough places for all the walk zone kids, only some could be admitted.

For the other half of the seats in a school, only the lottery established priority.

This division of the Boston schools into halves implicitly acknowledged a political reality. School choice divides parents into two "parties." People who live near good schools become the "walk-to-school party," while those who live elsewhere become the "school choice party." The priority policy in Boston (where people still recall the "busing wars" fought there a generation ago during desegregation) represented a compromise between the two, and the details of this compromise were adjusted from year to year based on which groups wielded the most influence.

Once adjustments in priorities and related matters had been made, the old Boston system, which is still used in many other cities, worked as follows. The central office asked families to list at least three schools in order of preference. The algorithm then placed as many children as possible into their first-choice school. When a school was the first choice of more students than it had places to accommodate them, students were admitted in priority order until all the places were filled. That is, the school immediately admitted the highest-priority students who had ranked it first, up to its capacity, and rejected the rest. This "immediate acceptance" algorithm then assigned as many remaining students as possible to their second choice. And then it moved to students' third choice, and so on. The central office assigned students who failed to get any of their choices to the nearest school with vacant seats.

At this point, you may be asking, What could possibly be wrong with a system that tries to give as many people as possible their first choice?

It all sounds benign, rational, and simple. *But being easy to describe isn't the same as being easy to navigate.* Like the old New York system, the Boston system presented families with difficult strategic choices and often made it unsafe for them to rank the schools in the order that reflected their actual preferences.

How so? Consider a child who lives within the walk zone of a school with a very popular half-day kindergarten. That school is her parents' second choice. Their first choice is a more distant full-day kindergarten that's almost as popular.

Because the parents know they have walk zone priority for the half-day kindergarten, they believe they can get it if they list it as their first choice. But if they list their true preferences, with the full-day program first and the half-day program second, they may fail to get either. They probably won't land a place in the full-day program because their child has no sibling or walk zone priority there. And in that case, their child would also fail to get a place in their second choice, which, since it is so popular, will have filled all its places with children whose parents listed it first. That is, such a popular school will have more first-choice applications than it has places, so it will immediately fill all its places when the assignment algorithm puts as many students as possible into their first-choice school.

And it gets worse. Those parents likely won't get their *third* choice either, since their child can only get in there if it isn't first filled by students who listed it as their first or second choice. This applies with even more force to any choices listed fourth, fifth, or further down the list. In fact, under the old Boston system, few people even took the time to rank more than three schools, since the odds of getting into a desirable school ranked lower than third were tiny.

Let's go back to our example. If the parents of the kindergartner list the full-day kindergarten first, the likely outcomes are either

1. they will get it (if they get lucky), or
2. they won't get *any* of their choices, and will be assigned to a school that is so unpopular it has vacant places after everyone's choices have been filled.

That's an educational Russian roulette. Either you win big or you lose bad. Unpopular schools are unpopular for a reason, so if an unpopular kindergarten is the only one these parents can get, they might decide to send their child to a private school (if they can) or even move to the suburbs. Keep in mind that BPS, as an arm of the Boston city government, has strong incentives to stop dissatisfied parents from opting out of the system. At best, such dissatisfaction causes unrest that can hurt the current government in the next election. At worst, unhappy parents will move out of Boston altogether, taking their tax revenues with them. For these reasons, effective access to good public schools is widely viewed by economists and urban planners as a key to keeping cities healthy.

Faced with complicated choices like this, many parents understandably played it safe. About 80 percent of the children were assigned to the school that their parents had *listed* first. On paper, the system looked hugely successful, with most participants appearing to have received their first choices. But the reality was very different: many of those parents were simply getting safe, strategic choices.

Strategizing like this can seem natural. I mentioned this before when we talked about unraveling, about having to make decisions early. When you're driving down a crowded street looking for a parking spot, you face this kind of decision. But think about it now as if you had to say which parking spot was your first choice — that is, if you applied for parking spaces the way Bostonians applied for school places, and the city would give as many people as possible their first choice. You see a vacant spot. Should you take it now (as you would if it *were* your first choice of places to park)? Or should you risk trying to find your *real* first choice, a spot right in front of where you're going—even if the odds of finding such a coveted spot are long?

If you had to get your parking spots through a clearinghouse that tried to give as many people as possible their first choice, you might want to designate as your first choice a spot that you knew you could get if you *said* it was your first choice, so as not to end up with something much worse.

Once again, even this simple choice is a *strategic* one, since you have to take into account other people's probable decisions, which will determine which spots are popular and likely to be occupied.

In 2003, reporter Gareth Cook wrote in the *Boston Globe* about an economics paper published earlier that year by Atila Abdulkadiroğlu and Tayfun Sönmez, in which they analyzed the Boston school choice system. Cook had no trouble finding parents who confirmed that the need to game the system was a major frustration. Said one such parent, "A lot of the alienation some parents have toward the choice system is solely attributable to the alienation of not making your first choice your first choice."

Later that year, the superintendent of Boston Public Schools and his staff invited Atila, Tayfun, Parag Pathak, and me to a meeting to talk about what we saw as the problems of the school choice system and how they might be fixed. Even getting invited to such a meeting involved matchmaking, a bit like arranging a blind date. Caroline Hoxby, an eminent education economist, had asked the dean of Harvard's Graduate School of Education to contact the BPS superintendent, Tom Payzant, to let him know that we would be worth talking to.

On the morning of October 9, the four of us took the "T" (the MBTA, Boston's mass transit system) to 26 Court Street, the Boston Public Schools headquarters. We'd already sent material to BPS describing how the school choice system might be reorganized to allow families to safely reveal their true preferences. But the staffers were skeptical, saying, "Maybe economics professors game the system, but ordinary families don't do anything so fancy." At that moment, I began to think that our first meeting would be our last.

But the mood changed as Tayfun began to speak about a laboratory experiment that he and a colleague, Yan Chen, had conducted. Economists increasingly use experiments to show how economic environments influence behavior. We build artificial economies in the lab, paying participants based on the outcomes they achieve. Experiments don't replace field observations; they complement them. The advantage is that in the lab, you can control and measure many aspects of the environment that can only be conjectured in the field. So in the lab, while you can't study the full range of complexities that actual parents bring to actual school choice, you can study very clearly whether the system used – in this case, the immediate acceptance algorithm – is a good way to allocate scarce resources.

In real life, when parents submitted preference lists of schools to BPS, no one could see their *true* preferences, only their *stated* preferences. But in an experiment in which artificial places were being allocated, an investigator could tell participants how much money they would earn if they ended up matched to a particular artificial school. That would enable the experimenter to compare the preferences that participants stated when asked to rank their choices to the payments that they would in fact have earned depending on which school they were assigned to.

The participants in the experiment didn't know that it was about Boston schools; as far as they were concerned, they were just trying to earn some money in the lab by trying to get as good an assignment as possible. And the experimenter could tell them which assignments were good for them, by telling them how much they would earn depending on what assignment they got. That meant that in the lab, the experimenter could know their true preferences, something that is much harder to know in the vastly more complicated environment of actual Boston schools.

In one part of Tayfun and Yan's experiment, participants playing Boston families tried to match up with schools using the existing BPS mechanism. The researchers paid participants $16 if they matched to their highest-paying school, $13 for the next-highest-paying, $11 for the third-highest-paying, and so on, down to $2 for the least profitable match. Since Tayfun and Yan determined the participants' preferences for each school in the experiment, they could see when anyone provided a rank order list that was different from his or her true preferences.

The BPS senior staffers were stunned by the results. Tayfun and Yan observed that some participants matched to their *stated* first choice, even though it wasn't the one that would have paid them the most. That is, participants intuitively understood that they might get better matches and be paid more for their participation if they stated as their first choice the best alternative that they could get —rather than trying to go for the full $16, and thus risk failing and making much less money.

After that, the BPS staffers were more open to our message.

I asked, "Which is the best kindergarten in Boston?" They suggested the Lyndon School in the West Roxbury neighborhood. I then asked, "Does everyone list the Lyndon as their first choice?" No, they said, that would be silly—you'd waste your first choice since you couldn't get into the Lyndon unless you had high priority. It was just too popular.

"Exactly," we replied.

Seeing the Problem

The upshot of that first meeting was that the Boston Public Schools senior staffers realized that there *might* be a problem with their current system. They invited us to prove to them that it was real and that it mattered.

Years later, after Tom Payzant had retired as superintendent, I asked him what that decision had looked like from his perspective. He replied that from the moment he took the job in 1995, he'd been concerned that his term might be dominated by the politics of school assignment—as in the 1970s and '80s when Boston schools had been subjected to court-ordered busing. When he heard that BPS could get experienced outside experts to help address the issues in a technical, nonpolitical manner, as New York had done, he was eager to try.

Over the next year, we dug into Boston's school choice data. We discovered not only that families had strong incentives to be careful about revealing their true preferences, but also that this decision had different consequences for different kinds of families. A parent who understood the old system but wanted to try to get her kid into a sought-after school would list it first but then play it safe by listing as her second choice a school that was unlikely to fill all its places with other people's first choices.

Parents who weren't careful in this way, either because they didn't know which schools were popular or because they didn't understand the system, often came to grief. About 20 percent of parents listed as their *second* choice a popular school that no one could get into except as a first choice. Their children often ended up unassigned to any of the schools on their list. In many cases, these kids would have been assigned if their parents hadn't made this mistake —if, for example, they'd listed their third choice as their second choice.

At least one parent group, in the school system's relatively affluent West Zone, devoted itself to compiling "intelligence" (for example, by talking to parents at local playgrounds) on how many children would be applying to the most coveted kindergartens and how many of those kids had older siblings already attending the same schools. These younger siblings would have the highest priority, so knowing their numbers would allow parents to estimate how many empty places would be available for other children—and thus the odds of their own children getting in. But gathering such information was an exhausting and error-prone process, and parents rightfully hated it.

Our findings ultimately convinced BPS that the choice system had to be changed. "A public school system's job is to provide equitable opportunities for all students," says Valerie Edwards, the senior BPS staffer who was crucial in initially convincing her colleagues that we were worth listening to. "The fact that our student assignment policy made it necessary for people to game the system meant it was a failed system. It meant that the public school system wasn't doing its job."

It turned out that the solution Boston adopted for its public schools was quite similar to the one we helped New York build for its high schools—and that these are both related to the clearinghouse through which most American doctors get their first jobs.

We're now ready to look at those solutions.

PART III

Design Inventions to Make Markets Smarter, Thicker, and Faster

8

The Match: Strong Medicine for New Doctors

THE SOLUTIONS TO problems in market design are sometimes invented, sometimes discovered, and often a bit of both. The designs for many markets have evolved, usually through trial and error, over the long span of human history. So we can sometimes discover a solution to a new market failure in a design pioneered in another market.

Such a solution still typically needs the invention of new modifications to fit the circumstances of the particular market in question.

A medical analogy may be helpful. Human beings are the product of an even longer evolution than are human markets. Our immune systems have evolved to help us fight diseases. But sometimes our immune systems fail, and the disease-causing germs win. What to do?

Well, nowadays, we can improve on our natural defenses with antibiotics. The first really important one was penicillin. Penicillin wasn't invented, it was discovered, by the Scottish immunologist Alexander Fleming in 1928. Fleming noticed that the *Penicillium* mold found on bread produced a substance that killed bacteria. That is, penicillin was a natural mechanism that the mold had evolved for dealing with bacteria. But penicillin didn't become available as a practical and widely available medicine until much more was learned about its medical properties; more productive strains of *Penicillium* mold were developed; and methods of industrial production were invented, with critical contributions by Howard Florey and Ernst Chain (they shared the Nobel Prize with Fleming in 1945).

Just as the *Penicillium* mold had evolved a way to deal with bacteria that could be adapted to fix a failure of human immune systems, ideas for correcting market failures can begin with an observation, "in the wild," about ways in which other markets have been organized.

Let's start with the market for new doctors that I mentioned in chapter 4. That market is particularly instructive, since at various points in its history, it suffered from many of the failures that are common in matching markets. So let's go back to the beginning, to understand the illness that the market for new doctors suffered and the cure that the doctors found for it. That solution proved to be the *Penicillium* mold for many matching markets.

A Cure for What Ails Doctors

Since about 1900, the first job American doctors take after graduating from medical school is called an *internship* or *residency,* in which they're supervised by more senior docs. Over the past century, it has become a requirement for getting a license to practice medicine independently. (Before 1900, doctors graduated from medical school and immediately began to "practice" medicine without further supervision.)

Residency jobs quickly became a major part of hospitals' labor force, a critical part of physicians' graduate education, and a substantial influence on their future careers. Needless to say, there's a lot of pressure on both sides to make good matches—on medical students to get good first jobs and on hospital residency programs to hire good young docs.

But almost from the start, something was amiss in the market for interns. An early symptom was that in their competition for scarce medical school graduates, hospitals began to try to hire interns a little earlier than competing hospitals. As a result, medical students were faced with decisions about internships earlier and earlier in their medical school careers. Students also were often forced to consider offers from one hospital at a time, without ever knowing what their prospects might be at other hospitals. This problem developed gradually, and you will recognize it as unraveling.

This unraveling of the market moved the date of appointment earlier, first slowly and then quickly, until by 1940 residents were sometimes hired almost two years before graduation. Just as picking Notre Dame to play in a bowl game was risky before the team had finished its regular season (see chapter 4), it was risky to hire a medical student two years before graduation. It was hard to tell who the good ones would be, especially since the first two years of med school are mostly spent in class, not with real patients.

As you can imagine, it was also hard for medical students to tell what kinds of jobs they might want two years hence. After getting an A in anatomy, a student might want to be a surgeon, only to discover in his third year, when he finally gets to scrub in to watch surgeries up close, that he faints at the sight of blood. But back in 1940, that would have been too late; he would have already been long since hired for the surgery internship he once thought he wanted. The student—and the surgeons who hired him—would both have made a bad match.

Although early hiring was pretty bad for both students and residency programs, we've already seen that unraveling isn't solved by self-control. Only in 1945, when a third party—medical schools—agreed not to release information about students before a specified date, was the timing of offers controlled. The medical schools embargoed transcripts, letters of reference, and even confirmation that students were enrolled. This certainly helped control the date of appointments: it's pretty risky to hire a student after his second year of medical school, but it's crazy to hire someone if you can't even confirm that person is actually a medical student.

Once the date of appointments was controlled—so that all hospitals were making offers starting at the same time—a new problem emerged. Hospitals now found that if some of the first offers they made were rejected after a period of deliberation, the candidates to whom they wished to make their next offers often had already accepted other positions. In other words, the wrong first offer put them out of the game, especially if it was rejected after any delay.

This, of course, led hospitals to begin making exploding offers. Now candidates had to reply immediately, even before they could learn what other offers might be available. That in turn led to a chaotic market that shortened in duration from year to year and resulted not only in missed agreements but also in broken ones. In other words, the market suffered from congestion: once hospitals learned that other hospitals were hiring fast, they couldn't take a leisurely approach to making their own offers. If they did, they wouldn't have time to make many offers before their favorite candidates were hired by others.

After having suffered from this kind of congestion for five years, the doctors did something extraordinary: they reorganized the market in a big way. Instead of a completely decentralized market, they proposed to organize the last stage of the market through a centralized marketplace, a kind of *clearinghouse*. It proved to be a critical, even historic, decision.

Under the new plan, third-year medical students would apply to residency programs on their own, as before, and the residency programs would invite them to interviews, also as before. But then a change: after the interviews had been conducted, the process of making offers would be done through the new centralized clearinghouse. This meant that students would submit to the clearinghouse a rank order list of the residency programs at which they had interviewed, indicating their first choice, second choice, and so on. In parallel, those residency programs would submit a rank order list of students.

Before the clearinghouse opened, both applicants and employers would exchange information about job descriptions (including wages and other features) and about the applicants' qualifications, so that each side could form well-considered preferences regarding the other. Note that they would do this *in advance*, so that the clearinghouse could use this information to suggest matches between applicants and jobs.

Part of the plan, as it was proposed to the participants, was an explanation of how these rank order lists would be processed to produce the suggested match. It's worth going into detail here—both about an initial, unsuccessful proposal and the later successful one—because these details are at the heart of market design. In this case, the first proposal raised critical questions about whether it would be *safe* (as we've seen, a critical factor) for participants to reveal their true preferences when they submitted their rank order lists.

In the initial proposal, students were asked to rank individual residency programs, while residency programs ranked students in clusters, with 1 being reserved for the most preferred students up to the number of available positions, 2 for the next most preferred group, and so on. The proposed algorithm first matched all residency programs and students that were each other's first choices (1-1 rankings). After that, residency programs matched students in their second group if those students had ranked the residency program first (2-1 rankings), followed by matches of residency programs' first choices with students' second choices (1-2 rankings), and so forth (2-2, 3-1, 3-2, 1-3, 2-3 . . .). The intention appears to have been to give an advantage to the students, since when the preferences conflicted, students' first choices were considered earlier than residency programs' first choices.

But after a trial run, students realized that it wasn't safe for them to confide their true preferences to the clearinghouse. Much as with the school choice systems decades later in New York and Boston, a student who listed as his first choice a residency program that didn't list him first might miss the chance of getting his second choice (even if he was that program's first choice).

One student who noticed this flaw in the proposed design was Hardy Hendren. He was preparing to graduate from Harvard Medical School in 1952, just as the clearinghouse was getting started. When he told me about it years later over lunch in Cambridge, Massachusetts, he had already retired (in 1998) from Boston Children's Hospital, where he had been chief of surgery. (His colleagues had given him the nickname "Hardly Human" for the long, complicated surgeries he was able to conduct.) Hardy entered the Navy during World War II, in 1943 when he was seventeen, and trained as a pilot before returning to college and medical school. As you can imagine, with that background, as he prepared to seek his first job as a doctor, he wasn't shy about expressing his concerns that the clearinghouse was unsafe for students.

Hardy also wasn't one to wait around for bureaucrats. And so, with a group of fellow students, he formed the National Student Internship Matching Committee, which organized opposition to the proposed algorithm. The committee recommended that it be replaced with a different way of processing the preference lists to determine a match; it became known as the Boston Pool Plan. This was, in fact, the algorithm that was finally implemented when the clearinghouse was used to match students and positions in 1952. It would prove to be a template, one strain of *Penicillium* mold, for a number of later market designs, including the redesign of the medical match itself that I was asked to undertake in 1995.

Back in 1952, this alternative clearinghouse was run successfully using card-sorting machines (there were few computers yet in use). What do I mean by "successfully"? Well, for one thing, lots of students and residency programs participated in 1952 by submitting rank order lists and then going ahead and signing the contracts suggested by the clearinghouse. I say "suggested" because all this was done in those days on a strictly voluntary basis. No one was forced to submit a rank order list, and no one who did so had to take the match that the clearinghouse proposed. Yet it wasn't long before "the Match" became an institution of the medical marketplace. It enjoyed very high participation rates for years without encountering further difficulties, in contrast to the troubled variety of market failures that had preceded this centralized clearinghouse.

When I studied this and other successful labor market clearinghouses, I discovered one of the secrets of their success. It was that they produced outcomes that were *stable*, in the sense that no applicant and residency program *not* matched with each other preferred each other to their assigned matches.

If a suggested matching isn't stable—that is, if there exists at least one applicant and employer who aren't matched to each other but would prefer to be—this unsatisfied pair is called a *blocking pair*. A matching is called *unstable* if there are any blocking pairs, since the members of a blocking pair can block the proposed unstable matching by instead making a match with each other. A stable outcome like that of the 1952 Match had no blocking pairs.

You can see why a clearinghouse that fails to produce stable matchings will have trouble attracting voluntary compliance with its suggestions. Suppose that some applicant and employer that aren't matched with each other *do* prefer each other to their assigned matches. That is, suppose they're a blocking pair. For example, suppose that the clearinghouse has suggested a match between the applicant and his third-choice employer and that one of his two preferred employers also prefers him to one of the people the clearinghouse has suggested for it. That applicant only has to make two phone calls—to the two employers he would prefer to be matched with—to determine that he is part of a blocking pair. The employer who prefers him will then have a reason to at least partially disregard the match suggested by the clearinghouse and to instead extend one of its offers to this applicant.

If this happens often enough, then in subsequent years the employer may withhold some or all of its positions from the clearinghouse, knowing that it can make better hires outside the system. (Remember how this happened in New York, when high school principals withheld positions from the old school assignment process.) Ultimately, if an algorithm produces unstable outcomes, there will be both applicants and residency programs that would rather be matched to each other than to accept the results of the Match. This creates incentives for these unhappy blocking pairs to circumvent the process.

Looking at it this way lets us see that a stable outcome is what we should expect from a very competitive market in which everyone is free to pursue his or her own goals vigorously. If there is a blocking pair—that is, a firm and a worker who want to be matched to each other—what's stopping them? If nothing is stopping them, we should expect the market not to yield an unstable outcome, since the blocking pair won't agree to it. But we've also seen in previous chapters that lots of things can stop such a pair from getting together: the market may be too thin, or too congested, or too risky for them to manage it.

Of course, so far that's just theory. But the theory that stable clearinghouses work better than unstable ones is supported by strong evidence.

For example, I found that when the British markets for medical interns experienced increasingly early appointments in the 1960s, each region of the British National Health Service devised its own centralized clearinghouse. Several used algorithms very similar to the one initially proposed for American doctors—the one that was rejected as unsafe for students. These unstable British clearinghouses failed and were abandoned after interested applicants and hospitals—blocking pairs—learned to circumvent them. In contrast, those British clearinghouses that produced stable outcomes succeeded and remained in use.

Back in 1952, economists hadn't yet figured out any of this, which makes Hardy Hendren's insight and his committee's grassroots efforts all the more impressive. The notion of stability wasn't clearly formulated until ten years later, in a 1962 article by David Gale and Lloyd Shapley with the intriguing title "College Admissions and the Stability of Marriage." The two authors didn't know about the Match, but they formulated an algorithm for finding stable matchings that I later discovered was equivalent to the one the doctors had adopted for the 1952 clearinghouse. Gale and Shapley called their version the *deferred acceptance algorithm*, and it eventually became the most important strain of the *Penicillium* mold for fixing failed matching markets — not least because they recognized that it always produces a stable matching, at least for markets without too many complications, such as couples who need two jobs in the same city. (But I'm getting ahead of myself.)

Lloyd Shapley was one of the founding giants of game theory. He wrote many papers that initiated whole areas of study, but this was the paper for which he was recognized with the Nobel Prize in Economics in 2012. David Gale would undoubtedly have shared it with Lloyd and me if he had still been living. Gale and Shapley weren't the first to discover the deferred acceptance algorithm, but they were the last: it was never lost again.

Here's how it works. I'll describe the algorithm as though applicants and employers are taking actions, but keep in mind that their only real action is submitting their preferences (step 0) — after which everything (step 1 on) happens in the computer, with no delay for decisions to be made and communicated.

- Step 0. Applicants and employers privately submit preferences to a clearinghouse in the form of rank order lists.
- Step 1. Each employer offers jobs to its top-choice candidates,

up to the number of its available positions. Each applicant looks at all the offers he or she has received, *tentatively* accepts the best one (the one highest on her preference list), and rejects any others (including any offers of jobs that were judged unacceptable and left off the applicant's rank order list).

. . .

- Step *n*. Each employer that had a job offer rejected in the previous step offers that job to its next choice, if one remains. Each applicant considers the offer he or she has been holding together with his or her new offer(s) and *tentatively* accepts the most preferred (highest-ranked) of these. The applicant rejects any remaining offers—including possibly the one that had been tentatively accepted earlier but is no longer the best offer received. (Note that applicants take no account of the step of the algorithm in which an offer was received; they look only at whether they prefer it to their other offers.)
- The algorithm ends when no offer is rejected, so that no firm wants to make any additional offers. At that point, each applicant and employer is (finally) matched by having each applicant accept whatever offer he or she had most recently tentatively accepted. That is, all acceptances are deferred until the end, when no more offers are forthcoming.

The astonishing thing that Gale and Shapley proved is that the final matching is always stable with respect to the preferences submitted by the employers and applicants, *whatever those preferences happen to be.* That is, when the algorithm ends and each applicant accepts whatever offer he or she did not reject (and any applicant who is not holding any offer is unmatched, as are any offers that are not being held), the resulting matching is stable. There aren't any blocking pairs, meaning that there aren't any applicants and employers that aren't matched to each other but that both wish they were.

How do we know this? (Get ready for a mathematical argument

so simple that it doesn't need any equations, just logical thinking. And it helped win a Nobel Prize.)

Suppose some applicant, call him Dr. Arrowsmith (A), and some employer, say the pediatrics residency program at Massachusetts General Hospital (P), aren't matched to each other. How do we know they wouldn't both prefer to be?

The key here is the word "both." It might be that A, who is matched to a residency program at the Rouncefield Clinic (R), would prefer to be employed by P (he ranked P before R on his preference list). But in that case, it must be that P never offered him a job during the algorithm, because if it had, he would have rejected R, which he never did since that is where he was matched. Why didn't P offer A a job? Because P filled all its positions with candidates it made offers to before it wanted to offer a job to A. That is, P filled all its positions with candidates it preferred to A. So if A would prefer to be matched to P, it must be that P doesn't return the favor. (That argument isn't hard to follow, but it's real mathematics, the kind that lets us understand something that isn't at all obvious.)

By going through this argument, we (you and I) just reproduced Gale and Shapley's surprising observation. We showed that for any doctor who prefers a different program than the one he is matched to, the program doesn't reciprocate his feelings. (That's equivalent to showing that for any program that prefers some doc to one of the ones it is matched to, the doc doesn't reciprocate.) Either of these facts demonstrates that the matching is stable; it has no blocking pairs.

This important bundle of simple ideas – a model of stable matching, the deferred acceptance algorithm, and the proof that it produces a stable matching for any preferences – was recognized with great fanfare – literal fanfares with actual trumpets – in Stockholm in 2012.

The Match succeeded as a marketplace because it solved the problems that had caused previous ways of organizing the market to fail.

It was attractive enough to students and hospitals to make the market thick, and it discouraged unraveling, since it was worth waiting for. It wasn't congested, since it asked for decisions in advance and executed the process of finding the outcome of all those decisions quickly. And it made it safe for doctors to reveal their true preferences, a point I'll explain in the next chapter.

Couples

The medical Match worked smoothly for decades. But it ran into trouble when women started to enter medical school in significant numbers.

Medical students are pretty busy, but one thing they can sometimes manage to do along with their studies is court their classmates. Starting in the 1970s, a small number of married couples appeared in the Match, and these couples were looking not for one internship but for two, near enough to each other so they could keep living together. This created a whole new problem for the Match, as these couples would sometimes decline the offers suggested by the Match. It wasn't long before some couples declined to participate in the Match at all, but instead communicated directly with hospitals that might hire them.

When this happened, couples sometimes found hospitals that preferred to employ them rather than hire the people with whom the hospitals had been matched. Having even a small number of couples arranging their first jobs outside the Match soon had a ripple effect that made this long-successful system also work less well for single students (who might find that the hospitals they were matched to didn't hire them) and residency programs (which started to notice that good candidates might be available before or after the Match).

The medical administrators responsible for the Match tried to make changes in their market design that would accommodate the needs of couples better. In the 1970s, these efforts took the form of having each member of the couple first be certified by the dean of their medical school as a "legitimate" couple and then specify one member of the couple as the "leading member." Each member would then submit a rank order list of job preferences, just as if he or she were single, but the leading member would go through the Match first. After he or she was matched to a job, the preference list of the other member would be edited to include only jobs in the same city.

Even when this process produced two jobs in the same city, couples sometimes didn't take those jobs. Instead, they worked the phones to find jobs they liked better, often with success. How come? I think of this as an illustration of what I call "the Iron Law of Marriage," which says that *you can't be happier than your spouse.*

How would this apply? Suppose some young couple gets two jobs in Boston, but one is a good job and the other not so good. The Iron Law says that couple had better get on the phone and see if they can find two good jobs somewhere else.

But if the Match was still producing a stable matching, couples wouldn't be able to find jobs that they preferred with employers that also preferred them. But they did—and that's where we can see how the market for new doctors had changed in a fundamental way when pairs of applicants started looking for two nearby jobs. The stability of the Match produced an outcome that made sure that a doctor and a job that weren't matched to each other didn't both prefer to be, but it couldn't generate a comparable outcome for two doctors and two jobs if those two doctors were a couple. The reason was the Iron Law: couples aren't like two job seekers who don't know each other; each member of a couple cares not only about the job he or she gets but also about the job his or her partner gets.

It turns out that there's no way the clearinghouse could produce an outcome that was also stable with respect to couples and jobs if it didn't allow those couples to express their preferences for *pairs* of jobs. But this is even more complicated than it looks. When I first studied this market in the 1980s, one of the things I showed was that even if you allow couples to indicate their preferences for pairs of jobs, for some preferences there may not exist *any stable matching.* I showed this by producing a *counterexample* for which no matching of applicants to employers was stable.

It became clear that the problem of designing the market when there were couples present, unlike the simple market of the 1950s or the simple model addressed by Gale and Shapley in 1962, was going to be a hard one. It was destined to become a more important problem as the percentage of women graduating from medical school climbed. (Today 50 percent of American medical students are women.)

Maybe that's why I still remember so very clearly the day in 1995 when my office phone rang, and I answered a call that would change my professional life. The call was from Bob Beran, the executive director of the National Resident Matching Program, as the Match had come to be called. The Match was in crisis—for a number of reasons, not specifically over how to handle couples—and Beran asked if I would agree to redesign it.

My first reaction when I understood what he was asking was, Why me? I knew why he had called me, of course: I had studied the Match, and stable matchings, and had shown that stability was empirically important for the success of a clearinghouse. I had even written a well-regarded 1990 book on matching, together with my friend Marilda Sotomayor. But I also knew that the only thing in our book that related directly to redesigning the Match were the counterexamples, such as the one about couples, which underscored that making the Match work well would be a tough task. I also knew that the simple mathematical conclusions for uncomplicated markets, such as the one reached by Gale and Shapley, weren't generally true when couples were in the market. I would be entering unknown territory.

That's why that call changed my professional life. Until I agreed to redesign the Match, most of my work had been theoretical. And as a theorist, it had been enough for me to point out that matching with couples was a hard problem. Now it would be *my* hard problem.

There were about a thousand people who had entered the Match that year as members of a couple (i.e., about five hundred couples — today there are almost twice as many). And it would be my job to find an elegant way of getting them the pairs of jobs they wanted, along with getting the rest of the graduating medical students and other applicants their preferred jobs. So I wasn't just a scientist anymore, with a responsibility only to try to understand better how things worked or failed. Now I was also committed to becoming, once again, an engineer, a market designer with a responsibility to try to make things work better.

As I recall, I set only one important condition for taking on the task: I wanted to collaborate with Elliott Peranson, who had been providing technical support for the Match for years. He is a remarkable, self-taught, practical market designer. He had more or less fallen into the job when he went to work for a consulting company that landed the contract to work on the Match. (In the years after that first contract, Elliott formed his own company that helped organize many other labor market clearinghouses.)

Over the years, Elliott had helped tweak the rules of the Match as new problems appeared and as the structure of medical education changed. I knew he was intimately familiar with what had worked and what had failed in past attempts. Indeed, it was Elliott who had helped change the Match from the original card-sorting procedure to a computerized process.

Elliott played a role that I have since found is essential whenever I've successfully helped design a complex market: he was the expert guide. As an economist approaching a new market, I'm something of a generalist, sort of like an experienced mountain climber approaching a new mountain. Even if I've studied the market from a distance, there are details I still have to learn, because details matter in design. I've already described in chapter 3 how Frank Delmonico and Mike Rees were our guides, and later our champions, in the design of kidney exchange. But Elliott was my first partner in design.

Over the next year, working together, we figured out how to handle not only single applicants and couples but also a number of other "match variations" that weren't gracefully handled by the deferred acceptance algorithm in its simplest form. (Some single doctors also needed to find two jobs to train for the specialties they wanted, and some hospitals needed flexibility to shift positions between different residency programs.) We knew we had to produce stable matchings whenever possible. We also knew that any acceptable algorithm that could handle couples wouldn't look just like the deferred acceptance algorithm; it would also have to track down and fix blocking pairs involving couples.

We eventually developed a hybrid algorithm that has come to be called the Roth-Peranson algorithm. It finds a preliminary matching of doctors to residency programs by starting with a deferred acceptance algorithm, which yields an outcome containing some blocking pairs. Then it tries to fix them one by one.

For reasons I'll return to when I talk about school choice clearinghouses, Elliott and I reversed the deferred acceptance algorithm that I've just described and used one in which the applicants applied for jobs, starting with the one they preferred most, rather than having employers offer jobs starting with offers to their most preferred candidates.

Even with all this work on algorithms, we knew we wouldn't be able to find a stable outcome if none existed. But to our happy surprise, when we looked at the data, it turned out it was almost always possible to find a stable matching. And this was true even though there were couples in the market who were now submitting lists that ranked pairs of positions in the order that the couple had decided it preferred.

Today dozens of labor market clearinghouses use our algorithm to help couples find jobs together, and it is virtually never the case that a stable outcome cannot be found. This proved to be a classic case of the engineering preceding the scientific understanding. In fact, only recently, my colleagues Fuhito Kojima, Parag Pathak, Itai Ashlagi, and others have helped explain why, in large markets without too many couples, we can expect that stable matchings will exist most of the time.

That the Match has been able to help hundreds of thousands of doctors, including tens of thousands of couples, find jobs in thousands of residency programs is a testimony to the flexibility of markets as a tool for coordinating complicated human endeavors. As medical education and employment have changed, and as the labor force has grown to include couples, it has been possible to adapt the underlying design of the Match so that it remains a marketplace that attracts participation.

But the same things that make markets attractive to participate in freely also put limits on what they can do.

Centralized Marketplaces Versus Central Planning

When I first undertook to redesign the Match, some of the medical administrators hoped that I would be able to help do something like central planning. In particular, one problem facing American health care is that it has always been hard for rural hospitals to hire young interns and residents, who prefer large urban hospitals where they can learn their trade by helping treat a big patient pool with diverse maladies, using the latest tools for diagnosis and treatment.

So one question I was asked was, Could I somehow tweak the Match to send more residents to rural hospitals that traditionally didn't fill all their available positions?

Long before I was asked to redesign the Match, when I proved a mathematical result now known as the Rural Hospitals Theorem, I had discovered that the answer to this question is *no*. It turns out that when a hospital doesn't fill all its positions at some stable outcome, it gets exactly the same set of doctors at every stable outcome. So as long as the Match had to function as a competitive market in which students and residency programs were free to match with each other if they both wanted to — that is, as long as the Match had to produce a stable outcome — we couldn't send more doctors to rural hospitals than wanted to go. Otherwise there would be some doctor at a rural hospital who was part of a blocking pair with another hospital that he preferred, and it would be more than a market could do to keep him there.

Notice that residents are paid, and how much they're paid is a factor in determining the desirability of jobs as reflected in their preferences. But wages are far from the most important influence on the desirability of those first jobs, which have a big influence on a physician's future career. This is why, for example, rural hospitals don't get more residents just by paying them more. It's cheaper for them to attract older physicians whose careers have already begun to take shape: even though those doctors have to be paid more than residents, they can do more and don't need the same close supervision by more senior doctors, and they aren't looking for a broad education as part of their compensation.

Education

While we're thinking about education, we'll turn in the next chapter to the problem of how to design school choice systems to assign children to public schools that their families would like them to go to. This is a problem for which my colleagues and I also helped design computerized clearinghouses based on the deferred acceptance algorithm. I'll describe how that made it safe for families to reveal their true preferences for which schools they would like their children to attend. That will also complete the explanation of why the medical Match works so well: not only does the Match create a thick market and avoid congestion, but it also makes it safe for doctors to reveal their true preferences for which jobs they hope to get.

Public schools are a matching market in which we don't allow prices to play any role at all in deciding which children get which schools. That's one way in which public schools are different from private schools: we don't let parents compete for public school places by offering to pay more for them, nor can public schools compete with other public schools for children by offering them discounts (they're all free). But we can bring some of the benefits of markets to public school choice by allowing families' preferences to play a role, much as the preferences of medical students and residency programs drive the market for new doctors.

Making this process work well is vitally important. A democracy that makes education both free and compulsory has an enormous collective responsibility to provide its youngest citizens with a top-quality education. That is hard to achieve, and educational successes and failures both have ramifications that last a lifetime. Allowing parents to have a say in which schools their children will attend is part of an effort to better match children, who have different needs, to available schools, which have different strengths. Just as the matching market for residents facilitates one of the most important transitions a doctor will ever make, so do the marketplaces for schools have a vast influence on kids' futures.

9

Back to School

THE PHONE CALL I received in 2003 from the New York City Department of Education was in many ways an echo of the 1995 call asking me to redesign the medical Match. Indeed, it was the success of the Match redesign that led Jeremy Lack to think of me when it became his job to deal with the failing New York City high school choice system. His intuition proved to be good: students in New York today enjoy (endure?) a high school choice process built on the same general principles as the medical Match.

Recall (from chapter 6) the problems that New York was experiencing when Parag Pathak, Atila Abdulkadiroğlu, and I began to look into the matter. It had a badly congested, paper-based choice system in which information was exchanged through the mail.

It wasn't even safe for students to reveal their true preferences, as some schools would admit only students who ranked the school as their first choice, and high school principals withheld places from the system to get preferred students later. So lots of students found their way to schools outside the official system. Its shortcomings had encouraged the rise of a black market of sorts, which was a clear symptom that it was failing. But the most pressing problem was the massive group of 30,000 students who couldn't be assigned to any school they'd chosen and had to be assigned administratively at the last minute.

Before we economists could give any advice, we knew we had a lot to learn about New York high schools and students. What we soon found validated something we'd already heard: many schools were *crowded,* and that created a lot of competition for the most desirable slots. Overall, there were roughly the same number of students and school places.

So why were there 30,000 kids who weren't receiving offers? One reason was the existence of another 17,000 kids getting multiple offers, who needed time to make and report their choices. The multiple offers were gumming up the system and making it congested. The most obvious first step would be to ensure that each child received only a single offer.

But before recommending that, we wanted to understand what happened under the old system when children received multiple offers. (We also were motivated by the fear of being remembered as the economists who had driven the last middle-class kids out of New York's public high schools.) We found that the overwhelming majority of students who were accepted at more than one school took the offer from the one they'd previously ranked highest. From that we concluded that these children wouldn't be harmed much if they received just the offer from that school. Meanwhile, that simple change would free up offers for other kids.

In addition, we concluded, if a single-offer system were to be "strategy-proof"—that is, if it enabled all students and parents to safely list schools in the order in which they liked them—it could help even those kids who, under the old system, would have received multiple offers. Why? Because it would allow them to safely list schools they had only a small chance of getting into—without losing out on the chance to be admitted to other schools they might like, such as Aviation High School.

Finally, we realized, for the new system to work well, it would have to encourage principals to reveal *all* their seats and not keep any in reserve. By withholding seats, principals had admitted students they liked better than some of the ones they would have received through the centralized process. Ideally, any new system would assure principals that they would like the students they got through the centralized process at least as much as those for whom they might have saved seats.

That's why, in the end, we suggested a computerized clearinghouse based on the deferred acceptance algorithm—that is, the same underlying algorithm that powers the medical Match. We believed its properties went a long way toward addressing New York City's problems, especially when organized with students applying to schools (rather than schools making offers to students).

In the new clearinghouse, the students submit a rank order list of the schools they prefer, and the schools compose a rank order list of the most desirable students (but now without seeing the students' lists). The first step of the new school choice algorithm begins with students applying to their first-choice schools. Schools reject those applicants who are in excess of their capacity, holding the applications of only their highest-ranked students. Rejected students then apply to their second-choice schools, and so on, with schools at each step keeping the applications of the highest-ranked applicants they can accommodate. All acceptances are deferred until there are no more rejections, at which point each school accepts the students whose applications it is holding.

Let's compare the old choice system with this new one.

Consider two fictional brothers, Amos and Zach. Amos applied to enter high school in 2003, the last year of the old system, while Zach applied to enter in 2004, the first year the new system was used.

Amos's first choice was Townsend Harris, in Queens, a selective high school that, under the old system, considered only applicants who ranked it first. Amos's second choice was another selective school, the Beacon School, in Manhattan, near his mom's office, which also considered only students who listed it first. His third choice was Cardozo, near his home in Queens. His fourth choice was Forest Hills, also in Queens. Amos understood that he was wasting a choice if he applied to both Townsend Harris and Beacon, since whichever one he didn't list first wouldn't even consider him. So he listed Townsend Harris first, Cardozo second, and Forest Hills third. He just missed getting into Townsend Harris and was admitted to Cardozo, his third choice, which he listed second. Because his grades were high, he was spared the anxiety and uncertainty of being wait-listed into the summer.

Zach, applying the next year, knew that the schools wouldn't see where he ranked them and couldn't penalize him for not ranking them first. So he listed his choices in the true order of his preferences, which were the same as his brother's: Townsend Harris, Beacon, Cardozo, and Forest Hills. (Just to be sure that he will be admitted somewhere, he listed a bunch more, although with his high grades, he wasn't too worried.) Once again, Townsend Harris got many more applications than it had places, and Zach, like Amos, just missed the cut. So at the next step of the new algorithm, he automatically applied to Beacon.

Beacon was also a very popular school, which under the old system used to receive about 1,300 applicants for its 150 places, so in the first step of the deferred acceptance algorithm, it rejected all but its top 150 applicants. But since acceptance was now deferred, Beacon didn't yet accept those kids who applied in step 1. When Zach applied, it compared him with the 150 who hadn't been rejected in step 1 and anyone else who applied in step 2, ranked them all, and rejected all but the top 150 of this new group.

Zach wasn't rejected in the second step nor in any subsequent step. And when the algorithm ended, he was accepted by Beacon. Unlike his brother, he could safely list his true preferences, putting Beacon second. That didn't interfere at all with his chance of getting admitted there after his rejection by Townsend Harris.

When students can list as many choices as they want, the deferred acceptance algorithm allows them to safely list schools in true order of preference; they won't lose a place just because someone else applied earlier in the algorithm. This works because *even if a student doesn't get into her first-choice school, she has just as much chance of getting into her second-choice school as if she had listed it first.*

The same thing is true for each choice; a student who doesn't get into his first seven choices has just as much chance of getting into his eighth choice as if he had listed it first. As long as students can rank as many schools as they like, their best strategy is their simplest: to rank schools in the order that they like them.

This is why Elliott Peranson and I flipped the medical Match algorithm to have students applying for jobs, and residency programs accepting or rejecting those applications, instead of the other way around. That way, we could assure the students that it was safe for them to reveal their true preferences to the clearinghouse. (It turns out that it's actually pretty safe for residency programs, and schools, to reveal their preferences truthfully, too, but that's another story, which is mathematically related to the fact that most people get the same match at every stable matching.)

For school choice, the fact that the deferred acceptance algorithm produces a stable outcome (one with no blocking pairs) also serves school principals well. To see why, imagine what might have happened if Zach had lobbied for admission to Townsend Harris after the algorithm had concluded. Could he have gotten in if his parents had stopped by the school and pleaded with the principal? Probably not, because if Zach preferred some school to the one he matched with, the school preferred every student it admitted to Zach. How do we know this? If Zach matched to his second-choice school, he'd already applied to his first choice and been rejected after it filled all its places with students it preferred. That's why it rejected his application.

Similarly, suppose some principal found that at the end of the assignment process, she wanted to admit many students whom she didn't get. Should she hope that their parents would drop by to ask if their kids could enroll? No. If those students had applied during the algorithm, their applications would have been accepted, since the school ranked them highly. The fact that they went elsewhere meant that when the algorithm stopped, they hadn't applied. That's because they had been accepted by schools that they liked better and had applied to earlier.

Therefore, when the algorithm finishes, there won't be any student and any school that aren't matched to each other but that would *both* prefer to be. For example, Zach liked Cardozo, but not as much as Beacon, which accepted him – so he wouldn't apply to Cardozo after being admitted to Beacon.

Notice that we've just repeated the logic that we used in the previous chapter to demonstrate Gale and Shapley's discovery that the final matching that results from the deferred acceptance algorithm is stable.

Details, Details

I made some simplifications in my explanation of how the deferred acceptance algorithm was adapted to fit New York school choice. It's worth mentioning a few of these simplifications, because details matter so much in market design.

Just as the medical Match had some special features (including couples looking for two jobs), so, too, does New York school choice. Also, school choice operates under a lot of constraints, and many people have to sign off on any innovations. Sometimes this led to unavoidable complications. Not all of these complications were unavoidable, but just as in kidney exchange, my economist colleagues and I were only advisers, and not all of our advice was adopted. (This is pretty typical of market design, by the way.)

So, for example, in practice the deferred acceptance algorithm is actually run more than once. That's because there are several specialized schools that form their preferences strictly on the basis of exam scores or auditions. By tradition, students offered a place in these schools must also be offered a place in one of the regular high schools. Thus members of this small group of students each receive two offers of admission even before the main round of the match is run. Their offers are determined by running the full deferred acceptance algorithm on all students' submitted preferences and then running it again for all the other students after these select students have been placed.

Another simplification I made in my description is that students can list as many schools as they like. We economists recommended that students be allowed to do just that, but on this important detail we did not prevail. So New York City students today can list only up to twelve programs among the hundreds that the city offers. Students who want to list more than that face a strategic choice of which twelve to list. But they still should list those twelve in order of their true preferences. That's perfectly safe; they can't do any better.

A more serious problem is that some students list too few choices to get matched. Each year the New York media report on students who listed only schools that require scores higher than they have. These students end up without school placements at the end of the main match. For them, there is a supplemental round, in which they submit a new rank order list of up to twelve schools from among those that still have seats. By that time, the most-sought-after schools have already been filled.

In 2011, after the main round of the match was announced, I received an email from "Jimmy," who said he was a thirteen-year-old student from Queens. He appealed for help because he'd been rejected by the five schools that he'd listed in the main match, despite solid grades. He told me that he dreamed of attending Harvard and was worried that he'd be choosing, in the supplemental match, among less desirable high schools that would limit his prospects. I couldn't help much—my colleagues and I may have designed the algorithm, but we have no role in its annual operation. But I did inquire of a former administrator about what might have gone wrong.

He immediately focused on Jimmy's math grade of 85 and said that none of the five schools that Jimmy had listed were likely to accept a student whose grade wasn't at least 90. Jimmy hadn't received good advice before he compiled his list.

I advised Jimmy to immediately talk to his middle school guidance counselor about how to approach the supplemental round. I ended with a little advice for when he applied to college—something I wished I could've have told him before he filled out his high school list: "Bear in mind that admission to Harvard and other top universities is very competitive, so be sure to apply to other schools, including some safe schools." Almost no one who lists twelve schools is left unmatched in the main round of the New York high school match. So if you know someone like Jimmy, encourage him or her to submit a long list of schools, just to be safe.

These small problems don't overshadow the benefits the new system brought to New York high schoolers. In the first year of operation, the number of students left to be matched to a school for which they hadn't indicated a preference was 3,000, down from 30,000 the previous year A more surprising (and equally satisfying) development was that in each of the first three years of operation, the number of students who got their first choice increased, as did those who got their second through fifth choices.

"It worked even better than we expected in terms of kids getting their top choices," Jeremy Lack says. "It really empowered the students."

We weren't surprised that the new system would immediately work better than the old one, but we'd made no changes in the algorithm in years two or three, so why did the system continue to improve?

Remember those seats that principals would withhold? It appears that principals were gaining confidence in the new system and understanding that they'd actually prefer students assigned by the algorithm to those they could admit later. As a result, more and more of them released all of their saved seats to the central match. It was as if, by creating a stable matching each year so that principals would be eager to enroll students through the centralized process, the Department of Education was creating thousands more places in desirable schools.

One reason that principals gained confidence was that DOE staffers did a good job communicating to them how the new system would work. Crucial in that effort was Neil Dorosin, the DOE's director of high school operations. The task of informing everyone about the new algorithm fell to Neil and his colleagues in the Office of Enrollment Services. Among those he had to educate was his ultimate boss, Chancellor Joel Klein.

"One day I got called down to talk to him," Neil recalls. "He was upset because he had a friend whose child didn't get into their first-choice school. The friend had a cousin whose child had gotten into the school, and it was their last choice. I had to explain why the system had to function that way" (i.e., to make it safe to list true preferences).

More than ten years later, New York's high school choice system is holding up well. The clearinghouse we designed is just a part of the sometimes forbidding gauntlet that families have to run to inform themselves about schools and decide how to rank them. But with the exception of some of the complications I mentioned, once families are informed, the school choice system no longer presents them with complicated strategic problems. Most important, it's no longer a congested process that leaves tens of thousands of students to be placed at the last minute into schools for which they've expressed no preference. (Even students who didn't get into the schools they hoped to attend have information and preferences about which less desirable schools they would like to go to.)

Boston

Our experience in New York prepared us well for Boston, although some of the problems there were different.

Boston Public Schools also decided to replace its old school choice system with one based on a deferred acceptance algorithm. Recall from chapter 7 that Boston already had a computerized choice system in place, in which parents submitted a rank order of schools, but it wasn't safe for them to submit their true preferences. The old Boston system used an immediate acceptance algorithm, in which schools immediately accepted the first students to apply, using the priority that each child had at each school only to break ties when more students applied than could be admitted.

The new deferred acceptance algorithm still gave priority to parents who lived in a school's walk zone or who had another child already attending a given school. And the algorithm was otherwise familiar to parents and administrators because it began in almost the same way as the old one had, with parents submitting a rank order list of as many schools as they liked. (In Boston, unlike in New York, there was no limit on how many schools can be ranked.) But now, instead of each school *immediately* accepting the highest-priority students who applied to the school as their first choice, schools *deferred* their acceptances until each could see if any higher-priority students applied later. They never denied a student admission until after the school had filled all its places with students who had higher priorities. This removed the strategic risk to which Boston's old immediate acceptance algorithm exposed families.

In the Boston context, in which students have priority at schools, imagine young Max, ready to go to kindergarten. Max has high priority at the half-day kindergarten across the street from his house, but his parents would prefer an equally popular full-day kindergarten at which they don't have priority.

Under the old system, if Max (or, more accurately, his parents) listed the across-the-street kindergarten as his second choice and then didn't get assigned to his first choice, he would have lost his priority in favor of students who listed the half-day kindergarten as their first choice. Today, with the new algorithm, if Max lists the half-day school second and isn't admitted to his first choice, he'll still be accepted to the kindergarten across the street if he has high enough priority.

In the new system, the school near Max doesn't fill its places in the first round. Even if it gets more applications than it has places, it waits until the very end of the algorithm before it accepts the applicants it hasn't rejected. This allows the school to wait and see who will apply and then accept the applications of the kids with the highest priority.

So now Max's parents can safely list their choices in the real order of their preference. If they don't get their first choice, they won't also be sacrificing the chance to get their second. And if they don't get that, they still have just as much chance to get their third choice, and so on.

This is the "strategy-proofness" that we also sought in New York: families no longer must think strategically about other people's preferences, which is what they were forced to do when they had to investigate which schools were popular.

Boston's school administrators, politicians, and residents embraced strategy-proofness, and the benefits of the new system emerged almost immediately for students entering kindergarten and sixth and ninth grades in September 2006. One sign of improvement was that families began submitting longer preference lists. Another was that children were getting matched to popular schools even when their parents didn't list those schools as their first choice.

Parents came to understand that they no longer needed to strategize; instead, they could simply list the schools they liked best. Now they could redirect their energy to figuring out which schools they preferred.

That's what Marie Zemler Wu and her husband, Sherman Wu, did to prepare for the choice process that placed their kids in school for the fall of 2011. The Wus, who live in Dorchester, wanted to find the right kindergarten for their daughter, Miryah. They began their search by meeting informally with about a dozen families in their neighborhood to compare notes on schools in the Boston school system's East Zone. "We got together four or five times to help each other out," Marie recalls. "We put chairs in the living room, and somebody would bring a bottle of wine. We were able to confirm impressions with each other. Everybody had roughly the same schools but in different orders."

For Marie and Sherman, what mattered most was a school with engaged teachers and a culture of parental involvement. A foreign language immersion program would be a bonus. The couple didn't strategize; they trusted the algorithm to do its work.

First on their list was the Henderson inclusion school, second was the Hernández K–8 Spanish immersion school, and third was Mather Elementary School. "We thought that unless we had the best lottery number, we'd be going to Mather," Marie says.

That's where the algorithm placed Miryah. Out of the group of twelve families that had gathered in Marie and Sherman's living room to compare notes, five ended up sending their kids to Mather. But they didn't have to list it first to avoid losing their priority there. "We're thrilled," Marie says.

Notice that school choice in Boston differs from that in New York in an important way. Boston schools don't really have preferences over students the way high school principals in New York do. The priorities that each student has at each school are assigned by Boston Public Schools. Those priorities are part of the design of Boston school choice, not something the design has to accommodate. And those priorities are set, and reset, in a political manner, with input from the school board, the city council, the mayor, and neighborhood groups. In fact, priorities and the menu of schools from which residents in different parts of the city can choose are regularly revisited in light of transportation costs and more explicitly political considerations. So the schools that parents can choose among, and the priorities assigned to particular children at each school, have changed since we helped Boston redesign its system. But the underlying choice algorithm is still used to assign children based on parents' rank order lists and the city's priorities.

That much of school choice is subject to regular small changes is probably as it should be (and is, in any event, unavoidable). Schools are a giant issue in city politics, and so it's natural that a lot of the moving parts should be adjusted as the school population changes, as neighborhoods change, and as political power shifts. But that still leaves a big role for the kind of market design that we economists do. Our goal was to give Boston, and other cities, a school choice design that would still work well even when they changed priorities, walk zone boundaries, and all the other political moving parts that are part of this intricate marketplace.

Spreading the Word

After our initial experience in New York and Boston, other school districts began to call us for help. Indeed, since founding the nonprofit Institute for Innovation in Public School Choice, Neil Dorosin has become the Johnny Appleseed of market design for school choice. With support from Atila, Parag, and me, IIPSC has helped design school choice mechanisms for Denver and New Orleans and has had input into school choice in Washington, D.C. As I write this, we have projects under way in several other cities as well.

Economists in Japan and Belgium also have begun to look into designing school choice systems there, and in England it seems to have become a priority for the Conservative Party.

In China, about 10 million students each year are assigned to colleges through a variety of centralized clearinghouses, a different one for each province. These all use a student's preferences and scores on a national exam as inputs to match each student to, at most, one college. For many years, the design of these clearinghouses gave students a risky choice, since if they didn't get into their first choice, they had much less chance of getting into any of their other top choices, much as we saw in Boston Public Schools. The current Chinese government is instituting some reforms to improve this college admissions process, including the redesign of the clearinghouses. It appears that in some provinces, the new clearinghouses have been modified incrementally, so that they now are somewhere in between immediate acceptance (as in Boston before we helped change the system there) and deferred acceptance (as in Boston and New York school choice today).

There's every reason to hope that in the coming decades, we'll be able to design even better school choice systems, although they may continue to rest on the same basic principles of making it safe and simple for families to participate and using preference information efficiently.

This matters deeply to me because schools play a critical role in some of the biggest issues facing our democracy, from income inequality to intergenerational mobility. We need to use schools better, so that our kids can get the education they need, whether it is provided at the closest school or not. School choice helps us deliver on the promises we make to all our children.

That being said, school choice systems, even if they are efficient, simple, and safe, don't solve the problems created by not having enough good schools. They are at best a Band-Aid applied to those persistent problems, by letting existing schools be used more efficiently. In a democracy committed to public education as a right, it's an open wound that we haven't figured out how to give every child a first-rate education. Schools in poor neighborhoods—even well-financed schools—are often bad schools, so poor children don't get the education that would help them pull themselves out of poverty.

That realization led to the decades-long experiment in forced busing that aimed to homogenize big-city classrooms and put everyone on an even educational keel. But court-ordered central planning ran aground (as most central planning does) because it is so difficult to force people to do things they don't want to do. (Rich and poor alike often want their kids to attend schools near home.) Forced busing gave many parents the incentive to flee the public school system, choosing instead private schools they felt better served their children. Sometimes parents fled Boston altogether.

And if there's one thing we've learned about flawed markets, it's that people flee from them, either physically or by resorting to back channels and black markets. Either way, flawed markets can undermine not just communities but whole nations. The Berlin Wall was a monument to that fact.

10

Signaling

As we saw in the previous chapter, markets can be dramatically improved when their design encourages people to communicate essential information they might otherwise have kept to themselves. But sometimes markets suffer from too much communication. It is a paradox of market design that *as communication gets easier and cheaper, it sometimes also gets less informative.*

We are seeing this ever more clearly as communication becomes more electronic. Email and social media are good examples. As the volume of messages grows, sorting the real ones from the spam, and the ones that deserve thoughtful replies from those that can be quickly acknowledged or ignored, becomes harder. And as entire markets move to the Internet, this overload of messages can make markets congested.

Take college applications: it has become much easier for students to apply to many colleges than it was a generation ago. Internet dating sites are another case in point: women with attractive pictures may receive so many messages that it becomes hard to tell which are worth responding to. The same can be said for job markets, in which so many applications can be made so easily that it's hard to distinguish qualified candidates from less qualified ones, and even harder to figure out which qualified candidates are actually interested in the job, and hence worth spending the time interviewing, assessing, and courting.

Notice that I've just spoken about signals for two quite different kinds of information. First, is the candidate *qualified* enough for the college, the romantic partner, the employer, to be worth further investigation? And second, is she *interested* enough in them to repay that effort? Both kinds of information are especially valuable in congested markets, as there isn't time to explore every possibility. So signals, and how to send them, can be an integral part of a market's design.

College Admissions

Let's start with college admissions, since applying to college has always involved quite a bit of communication — of high school grades and teacher evaluations, and of personal information. It will give us an idea of how easier communication can lead to congestion, and why signals of interest are important.

Not so very long ago, each college required its own proprietary application. Today many American colleges accept applications over the Internet through a centralized site called the Common Application, in which all the materials can be assembled online and the same application essays can often be used to apply to multiple colleges. More than five hundred colleges accept applications through the Common App.

It's easy for students to apply to many more colleges than they used to — and they do. This in turn makes it hard for colleges to assess just how interested a given applicant actually is. So colleges look for other signals. A high school student who visits any college campus should make sure to sign the guest book in the admissions office, because that's one of the signals.

You might ask, Why do so many colleges use the Common App instead of requiring individual applications? Individual applications would make it harder for students to apply, so each application received would be a stronger signal of interest. The answer is that while the very top schools can do what they like, all the others must face the fact that the Common App has become a very thick marketplace. It attracts so many applicants that a college that now insists on individual applications might not get enough applicants, since many students are drawn to the convenience of using the Common App.

Instead, colleges often insist on a supplementary essay, different from those that can be sent to other colleges. That way, they can not only get another writing sample but, more important, can filter out applications from students who aren't interested enough in them to write that extra essay.

In South Korea and Japan, colleges make sure that students who apply to them can't apply to many other colleges. Colleges make applicants take admissions exams, and many colleges limit the number of other colleges to which students can apply by scheduling their exams on the same day as competing colleges. Thus any student who takes a particular college's exam is sending a very strong signal of interest.

American colleges do something similar when they admit students through "early admissions" programs. Most applications are made in January of a high school student's senior year. But many colleges also accept early applications in November. Quite a few of these colleges offer "binding early admission," in which they impose an additional requirement that students commit to attend the college if accepted.

But even colleges that don't place that requirement on students still admit many of their students under "early decision" programs that require applicants to declare that they are applying early to just one college. These commitments are enforced in part through the cooperation of the high schools, which have to send recommendations and a transcript of grades to any college to which the student applies. The high schools have an interest in making sure that their students live up to their commitments. After all, they will have other students applying to the same colleges in future years and want those later students' early applications also to be treated as a strong signal of interest.

That strong signal is one of the reasons that colleges admit a higher percentage of applicants from their early pool than from their regular pool. So if your high school student is organized early enough in the year, applying to one college early—even without committing to go there if accepted—will increase his or her chance of receiving a letter of acceptance, while still leaving open the opportunity to apply elsewhere during the regular admissions season.

One consequence of the Common App is that colleges need to admit more students to end up with an entering class that is the size they want, since it's now so easy for students to apply to a college without any deep desire to attend that college. Perversely, however, overall acceptance rates—especially for applications that aren't made early to a single college—have gone down, and each individual application is less likely to be accepted, because even though the absolute number of acceptances is going up, the total number of applicants is going up even faster—thus the ratio falls. That in turn prompts students to submit even more applications, which makes acceptance rates still lower and makes it more important for each student to submit even more applications. So the Common App makes it easier to apply to multiple colleges, but it also contributes to a vicious circle in which applicants need to apply to more colleges than before.

Of course, a college application signals more than the applicant's interest in the college. It also transmits many signals about why the college should be interested in the applicant.

Exam scores signal something about aptitude, as do high school grades, which also signal something about study habits. Elective courses and extracurricular activities signal an applicant's talents, skills, and interests. As colleges pay more attention to these signals in order to help them sort through increasing numbers of applications, the signals themselves may take on different meanings. That is, signals sent deliberately may convey different information than signals sent in passing. For example, does the applicant play the trombone because he likes trombone music better than harp music or because someone told him that colleges look for students who might join the marching band? Was the lacrosse enthusiast drawn to that sport by its speed and finesse, or did she hear that college coaches could influence admissions? Shaping high school activities to colleges' demands is not necessarily a bad thing; it's a lot like choosing your college major based on what employers value. But it does change the meaning of the signal.

Signaling for a Job

When it's time to apply for a job, whether a candidate went to college and what he or she did there comprise a powerful mix of signals about interests and skills and talents. That's one reason so many jobs demand a college degree, even if the job itself has little to do with what is taught in college. Going to college and doing well signals not only what the job applicant may have learned but also that he or she is good at learning things. That's a valuable skill in itself for just about any demanding job.

For an employer looking to hire young auto mechanics, however, college might not be the best signal, even though auto mechanics also need to be good at learning things. The employer might prefer someone who has spent his or her teenage years turning old cars into hot rods using parts salvaged from junkyards. That would signal interests and skills that would be handy for a car mechanic.

Applicants need to signal their qualifications and interest even for jobs that require very specialized education and long training, despite the fact that such preparation is itself a strong signal. Suppose you want to be a professional economist and do the kinds of work for which you need the skills acquired while earning a Ph.D. You'll be entering into quite a specialized labor market, one of many that exist for people with advanced training in various fields.

As a new economist with a Ph.D., you'll be competing with about 2,000 other new economists graduating each year from American universities, as well as others from overseas who want jobs in perhaps five hundred American colleges and universities, quite a few foreign universities, government and international organizations, big banks, and, increasingly, companies such as Google and Amazon whose business involves making markets.

At the heart of this marketplace is a three-day meeting each January organized by the American Economic Association, during which recruiting committees from university economics departments and other potential employers interview job candidates. Each new Ph.D. sends out lots of applications—I've seen people submit nearly a hundred—and so even a department with a single available position can expect to receive hundreds of applications.

This trend has been building for decades, and in recent years it has escalated because most applications can now be submitted online. The market has become congested. There's no way an economics department that has received hundreds of applications can interview every applicant in these three days.

If you didn't know how competitive this market is, you might think that at the end of evaluating the application materials, recruiting committees would interview the twenty or so candidates they liked best. In fact, that strategy works well for Harvard and Stanford and a handful of the most prestigious employers. But before I moved to Harvard, I worked at the University of Pittsburgh, a fine institution that isn't, however, at the very pinnacle of American universities. There, a strategy of interviewing the top twenty applicants wouldn't have worked.

Here's why. If the University of Pittsburgh had time to interview only twenty people at the January meeting and the recruiters interviewed the ones they liked best, all of those candidates might ultimately accept better offers somewhere else, and all the recruiters' efforts would be wasted.

So most economics departments need to choose a portfolio of candidates to interview—taking into account not only how promising the candidate looks to them but also how likely they are to be able to hire that candidate. Making that assessment isn't easy, and candidates regularly fall through the cracks—not getting offers from any of the departments that interviewed them, even as other departments that might have hired them chose not to interview them.

To improve this situation, a committee that I chaired developed and implemented a "signaling mechanism." In December, after most job ads have appeared, candidates are invited to log on to an AEA website to send no more than two signals of interest to departments to which they have already applied. Departments receive these signals and know that candidates chose to use one of their signals to indicate that they'd like to be interviewed for the advertised job.

Signals like these would have been mighty helpful back when I was at Pitt. Although we knew it didn't make sense for us to interview *only* the most prestigious candidates, we certainly didn't mind interviewing *some* of those candidates, since there was always the chance that their other interviews might not lead to offers. It would have helped to know which top candidates might have been seriously interested in us as opposed to being just tire kickers.

Signaling for Love

Something very similar happens at many Internet dating sites. Attractive women get more emails than they can answer. The men, who find that many of their emails go unanswered, react by sending more emails. Moreover, these emails become less informative, because the men submitting them are less likely to study the information contained in each woman's profile and how best to approach her. The women in turn reply to a smaller and smaller percentage of the messages they get, and the men respond by sending even more, and even more superficial, messages.

Economists call such superficial messages "cheap talk." When talk is cheap, it doesn't reliably signal anything. An email whose subject line is "I love you" means little when it is sent to many recipients. That helps explain why expensive diamond rings often accompany proposals of marriage (and why wearing them signals to other potential suitors that the wearer is uninterested in further proposals, and is thus unlikely to be worth pursuing).

My colleague Muriel Niederle, who was on the committee that designed the signaling mechanism for economists, was curious whether the same kind of mechanism might be helpful to an online dating site. She and our colleague Soohyung Lee introduced a signaling mechanism into a special event held by a Korean online dating/marriage site. The event gave men and women the ability to send a contact message—a "proposal"—to as many as ten potential dates over a period of five days. Each participant also got two virtual "roses," each of which they could attach to one of their proposals as a signal of particular interest. A randomly selected 20 percent of participants were given six additional roses, so that this group had the advantage of being able to send a signal of interest to more people.

After the initial contact period, participants could decide which, if any, of the proposals to accept. Men and women who had accepted each other's proposals were given mutual contact information. The experiment allowed Niederle and Lee to see the effect of roses on the acceptance rate of proposers, as well as the relative success of participants who had many roses compared to those who had few.

This site also offered other services, and as part of its algorithm for suggesting matches, it rated everyone on his or her desirability as a marriage partner, using a measure that took into account judgments of physical attractiveness and also verified financial, employment, education, and family data. Participants were not told their own desirability score or anyone else's. Lee and Niederle were able to examine the effect of proposals, with and without roses, as a function of the relative rated desirability of both the sender and the receiver. They classified participants as being in the top, middle, or bottom part of the distribution of rated attractiveness.

It turned out that proposals with roses attached were 20 percent more likely to be accepted. According to Lee and Niederle, "This positive effect of sending a rose is comparable to (and about three-quarters of) the benefit of being in the middle desirability group relative to being in the bottom group." That is, a proposal with a rose attached was regarded much the same way as a proposal from a more desirable potential mate. And the effect of a rose was clearest when the sender was in a higher desirability group than the receiver.

Finally, participants who had more roses to send did better than comparable participants who had only two roses to attach to their proposals. So "saying it with flowers" helped proposals stand out: this signal was effective.

Signals Everywhere

In job markets, as in dating, there are lots of ways to signal interest. The Internet has introduced some new ones, but humans have also spent tens of thousands of years developing ways to send signals of interest that are easy to interpret. Interestingly, many of the signals that are easiest to interpret are also, in some sense, the costliest to send.

Biologists believe that evolution sometimes leads to costly signals, such as the peacock's tail, that assist in the mating market. That is, a male peacock doesn't get a lot of direct benefit from a huge, heavy, colorful tail that advertises his presence to predators and makes it harder to escape from them. But it is a powerful signal of how healthy he is, since a less fit peacock couldn't produce such a large tail or would have been eaten by foxes because he was too slow to fly to safety. Thus a beautiful tail advertises to the local peahens that this is a peacock with impressive genetic resources. (Evolutionary biologists refer to the "four Fs" of natural selection: feeding, fighting, fleeing, and, um, reproduction. A large tail is a "handicap" in the first three categories, which sends a signal about his underlying fitness, and that increases his opportunities in the fourth.)

"Peacock tails" aren't just found in the animal world. Before 1994, when banks were allowed to open lots of branches, main bank buildings were imposing structures, with big marble lobbies from which you could sometimes see, behind bars, a massive vault. What was going on with that? Well, banks hold your money, and to signal that they were reliable places where it would be safe for you to leave your money, they wanted you to know that they had ample resources and weren't going anywhere. An impressive lobby was their peacock tail: a less well-funded bank couldn't afford to build such an expensive building, or might have built a building that could be more easily turned into a restaurant in case the bank failed.

The value of costly signals is why many colleges pay special attention to high school students who take the trouble to make campus visits. It's cheap to apply to a slew of colleges, so a mere application doesn't necessarily signal a deep interest. An actual visit, which is time-consuming and possibly expensive, is a stronger signal.

A signal of *interest* is different from the other kinds of costly signals that colleges value and not everyone can send, namely high grades and test scores. Good grades signal good study skills and hard work, high intelligence and aptitude, or all of the above, which are all attributes that help students succeed in college. Signals such as high grades—or a peacock's tail or big bank building—don't indicate interest; they are marks of *desirability.* That is, the gaudy tail doesn't signal how much the peacock is interested in a particular peahen; it signals how much the hens should be attracted to that particular cock.

Markets work best when they allow both kinds of information to be reliably transmitted. As we've seen, in a congested market—one in which it's impossible to explore every opportunity—*it helps to be able to signal not only how desirable you are but also how interested.* That's why, while many of us might wish to marry a movie star, we devote most of our efforts to finding and courting more realistic mates who might also like to marry us. (*Mutual* interest is what separates courting couples from stalker and prey.)

Asking a person out on a date in person offers lots of opportunities to send both kinds of signals. By comparison, arranging a date on the Internet, which makes the dating market thicker by making initial contacts easier, also makes it harder to send credible signals to cut through the congestion. Of course, some signals of desirability can be transmitted even over the Internet through pictures and credentials: how you look, where you went to school, what kind of job you have, and what hobbies you pursue are all signals that help someone decide how interested they might be in you.

In person, you can also send costly signals of your interest. Flowers and other forms of attention, from remembering birthdays to sending Valentine's Day cards, signal that you're devoting your attention—a valuable, costly good in itself—to the person you are courting, and so it might be worth that person's while to devote his or her attention to you. That's why mechanisms that limit the number of signals you send can accomplish the same thing over the Internet. When other signals may be cheap talk, these signals indicate that you are interested enough to use scarce resources that you can't just send to everyone. So a scarce signal isn't cheap talk; it comes with an opportunity cost—you could have sent that signal to someone else instead.

In labor markets, a cover letter in a job application can provide a powerful signal of interest, especially if it shows that the candidate has spent time to learn about the job for which he or she is applying, or even that the applicant has spent time carefully crafting a letter addressed specifically to the job in question. But trying to fake a signal—for instance, by mass-producing what is meant to be taken as an individually crafted letter—can be costly if detected.

My son Aaron, who is a professor of computer science, encountered just such a careless signal when he was on the admissions committee at Carnegie Mellon University. One Ph.D. applicant submitted a passionate letter about why he wanted to study at CMU, writing that he regarded CMU as the best computer science department in the world, that the CMU faculty was best equipped to help him pursue his research interests, and so on. But the final sentence of the letter gave the game away:

I will certainly attend CMU if adCMUted.

It was proof that the applicant had merely taken the application letter he had written to MIT and done a search-and-replace with "CMU" . . . and hadn't even taken the time to reread it! Had he done so, he would have noticed that every occurrence of those three letters had been replaced.

Auctions as Signals

Notice that, in the preceding examples, you may have to send costly signals even though those costs are partly wasted. For instance, if you go to college and study hard to signal that you're good at learning and then take a job for which what you learned in college isn't helpful, you paid a heavy cost that doesn't directly help your employer (although it may give you big private benefits apart from sending a signal).

If there were a cheaper way for you to send an equally convincing signal, your employer would have been just as happy to use that signal to sort through applicants. But if there were a cheaper, less time-consuming, easier signal to send, maybe people who aren't as good at learning as you are (and wouldn't have enjoyed college as much) would be able to send it, too, and it would become less informative.

That's why restaurants don't just rely on ads to signal how tasty their food is, since any eating establishment can advertise that it serves good food. It's also part of the reason restaurants sometimes have prices low enough that long lines form outside. Waiting is costly for customers; they could be doing something else, or eating sooner. The cost spent waiting doesn't directly turn into profits for the restaurant, and some of the people who are waiting would be happier to pay a little more for their meal and not have to wait.

So why doesn't the restaurant raise its prices and eliminate the line? That is, why does it forgo some immediate extra revenue that higher prices could bring? Because that long line sends a signal that the restaurant across the street with empty tables can't easily mimic—that is, a lot of people think this is a good restaurant, worth waiting for, and if you haven't tried it, maybe you should get at the end of the line instead of going across the street.

Aside from the signal value of a long line, the restaurant doesn't benefit from the time that patrons waste while waiting, just as the peahen doesn't benefit from the peacock's tail, or the bank customer from the high-ceilinged lobby. But there's an ancient method of signaling in which the cost of the signals to the signaler is exactly equal to the benefit to the recipient. I'm speaking of auctions, in which the high bid not only signals how goods should be allocated but also pays the seller of the goods.

Suppose someone is selling a painting by Rembrandt. One reason that auctions are such ancient and useful ways to sell something is that the high bidder sends a costly signal that he values the painting more than any of the other bidders, and the cost of his signal isn't wasted at all: the seller and the auctioneer receive the money that the high bidder spends. So the winning bid isn't *just* a costly signal (although it is that); it's also an efficient, direct transfer of wealth to the owner of the painting from the bidder who values it the most and signals with his high bid that he should be the one to get it.

There are lots of ways to run an auction; auction design is one of the most active parts of market design, as well as one of the oldest.

Goods ranging from works of art to cattle are often sold in "ascending bid" auctions, in which the auctioneer calls out progressively higher prices, until only one bidder remains, and that bidder pays the last, highest price called by the auctioneer. Sometimes items are sold instead in "sealed bid" auctions: each bidder submits a bid without hearing the other bids, the bids are all opened at the same time, and the highest bidder wins, sometimes paying the amount of his bid and sometimes paying the amount of the second-highest bid.

Paying the second-highest bid may sound odd, until you notice that in an ascending bid auction, the winning bidder pays the price at which the second-highest bidder dropped out. So in both an ascending bid auction and a second-price sealed bid auction, the highest bidder gets the object at the price just beyond what the second-highest bidder was willing to pay. Both of those auction formats make it easy to decide how much to bid, if you know how much the object is worth to you. That's because if you think of the winning bidder's profit as what the object is worth to him minus what he has to pay for it (and each losing bidder's profit as zero), it's perfectly safe for bidders to bid the object's full true value to them in a sealed bid auction, or to stay in an ascending bid auction until the auctioneer reaches the full amount they are willing to pay. Win or lose, a bidder can't make a higher profit by bidding something else.

That isn't obvious at all, but if you think about it carefully, you'll see why it's true. Consider the second-price sealed bid auction, in which the high bidder receives the object and pays the second-highest bid, while the other bidders pay nothing and receive nothing. By bidding less than the object's true value, a bidder sometimes turns a profitable winning bid into a losing one, and by bidding more than the true value, he sometimes turns a losing bid into an unprofitable winning bid at which he pays more than the object was worth to him.

Let's look at that situation carefully. Suppose your true value for the object you are bidding on is $100. If you bid $100, either your bid is the highest, in which case you win the object and pay the amount of the second-highest bid, say $90, or someone else bids more, in which case you pay zero and get nothing.

If your bid is the highest, you get an object worth $100 to you for only $90, so you make a profit of $10. What will happen if you bid $95 instead of your true value? You will still pay $90, since it's a second-price auction, so you will still make the same profit, $10. But suppose you bid even less, say $85? In that case, you won't be the high bidder, and so you will earn zero profit. Therefore, if your true value is higher than the other bids, lowering your bid below that value doesn't help you when you remain the highest bidder, and if you lower your bid so much that you're no longer the winner of the auction, it hurts you; your profit drops to zero.

Suppose instead that your true value, $100, is lower than what someone else bids. Suppose the highest bid submitted is $120. If you raise your bid to something above $100 but still below $120, it won't help you, since you still won't win anything, or pay anything. But if you raise your bid to above $120, you will win the auction and pay $120 (now the second-highest bid) for something that is worth only $100 to you. Bad move: you've converted a profit of zero into a loss of $20.

So it's safe to bid your true value in a second-price auction of this kind, since you can't do better by bidding anything else.

Notice that while a second-price auction makes it safe for bidders to bid the true value to them, it doesn't necessarily impose a cost on the seller, even though the seller receives only the amount of the second-highest bid. That's because in a first-price sealed bid auction, for example, it *isn't* safe for bidders to bid their true value; they have to bid less than that if they are going to make any profit, since if they win the auction, they will have to pay the full amount of their bid. So the seller in a first-price auction receives the amount of the highest bid, which is, however, less than the true value of the highest bidder. By comparison, in a second-price auction, the seller receives only the second-highest bid, but the bids are higher. That is, when the rules of the auction change, the bids change, too. In fact, there are reasons to think that these two effects balance out.

The situation changes when you don't know how much the object is worth to you. Suppose an oil company is bidding on the right to drill in some location. Its geologists estimate how much oil is under the ground there, but it's only an estimate. Other bidders have their own estimates, some of which might be more accurate, some less. In any case, the oil company might get a signal of how much it should be willing to pay by hearing the bids of the other bidders, which would convey some information about those other companies' estimates of the amount of oil available.

In this environment, an ascending bid auction is different from a sealed bid auction, even a second-price auction, since when the bids are sealed, bidders can't learn anything from how the other bidders behave. But in an ascending bid auction, when you see other bidders drop out, you know their estimates of the value aren't as high as yours. That might tell you that your own estimate is unrealistic: if there was as much oil in the ground as your geologists estimated, other companies should have seen it, too.

In contrast, a sealed bid auction, in which you can't see when the other bidders drop out, might make it risky to bid at all, since a company with an unrealistically high estimate of how much recoverable oil is in the ground might suffer the "winner's curse"—that is, win the auction only because it overestimated the value of winning and paid too much.

But first-price auctions, in which the winning bidder pays what she bids, have their own charms and exist in many varieties. One version of a first-price auction is used to sell cut flowers in bulk, in a "descending bid" auction. The auctioneer sets up a "clock" that has the current bid on it, starting with a very high bid and quickly descending, until some bidder stops the clock by offering the price it currently shows, which is higher than any of the other bidders have offered to pay, as they haven't already stopped the clock.

Since the first bid stops the clock, these auctions can be very fast — a good thing, since time is of the essence when you're buying cut flowers. A big international marketplace operates this way in the Netherlands, right near the Amsterdam airport, from which flowers can be shipped around the world. As a result, this kind of descending bid auction is often called a Dutch auction.

Most of the signals we've talked about so far are signals that people send about themselves. College applicants, job candidates, and prospective mates signal their talents, skills, and interests. You could think about all those signals as being sent *from the seller to the buyer.* Signals about quality are like that: they're of the form, *I'm a good student, a desirable mate, a restaurant that people are willing to stand in line for.* Signals about interest are also from seller to buyer: they're of the form, *I'm really interested in working for you, in attending your college, in dating you.*

But when, in 1993, Congress decided that the federal government would sell licenses for businesses to use radio spectrum instead of just giving those licenses away for free to firms with powerful lobbies (such as radio and television broadcasters), it needed some signals from the *buyers* about what the best uses of that increasingly scarce resource should be. In this case, the seller — the government — knew less about the value of what it was selling than did the potential buyers.

That's often the case. Experts can estimate the value of that Rembrandt painting, for example, but its true value is unknown until the bidding stops. That's why one of the ancient uses of auctions is called *price discovery:* letting the market tell you what price you can get for what you are selling, and to whom you should sell it to get that price. Auctions are matching markets that match sellers with the buyers who most value what is being sold.

But spectrum is more complicated than a Rembrandt painting, since it can be divided and combined in lots of ways for different uses. When Congress ruled that the Federal Communications

Commission should auction off spectrum licenses, it specified that the goal would be selling those licenses in a way that allocated them to the most valuable uses. The FCC needed an auction format that was flexible enough that businesses that wanted to use spectrum in different ways would be able to bid for what they wanted.

For example, some bidders might want to assemble a set of licenses that would let them run a nationwide cell phone network. Cell phones require only a relatively narrow band of frequency, so a cell phone provider would be looking for licenses to broadcast narrowly, but all over the country. Other businesses might need to assemble a broad band of frequencies in whatever geographical areas they wished to serve, so that, for example, their subscribers could download movies. They would want to license a lot of radio frequency, but maybe just in a single city.

Note that a business plan for using radio spectrum depends on assembling a *package* of licenses, much as a real estate developer who wants to build a building with a large footprint in a crowded city might have to assemble a package of land parcels. The whole package—of spectrum or of land—might be much more valuable than the sum of its parts. Just as a builder couldn't construct a big building if he failed to assemble a connected plot of land, an Internet provider couldn't provide broadband service if it didn't assemble a package of licenses covering a broad spectrum, or a cell phone provider couldn't provide service unless it assembled a package covering a wide area.

The FCC and the economists advising them and potential bidders quickly realized that selling spectrum licenses one at a time would be bad market design, as it would make it too risky for bidders to assemble the packages they needed. That is, if licenses were auctioned one at a time, bidders would have to bid very cautiously for fear of being left with only part of the package they wanted, which would be less valuable to them than what they had paid. To put it another way, it wouldn't be safe for bidders to fearlessly bid for each license the amount it would be worth to them as part of the right package, since they would have to pay their bids even if they couldn't ultimately put that package together.

To address this problem, the FCC adopted an auction design in which many licenses were sold at the same time, in "simultaneous ascending" auctions, with the rule that no auction would close until they all closed. That is, right up to the end of the auction, bidders could adjust the package of licenses they were bidding on, since the auction for each license would remain open until there was no more bidding on any license.

This didn't make it completely safe for bidders to assemble packages — they still faced a risk that they would be the high bidder on many licenses but that the price of the remaining licenses they needed would become too high for them to complete the package. But it went a long way toward solving that problem, because most packages allow for a certain amount of substitution; so as some parts of the package became too expensive, bidders could shift their bids to a different package. And taken all together, the bids themselves determined how the mix of winning packages should be divided among competing uses to create the most value.

The auction design also had another problem to solve: for the market to do its job, bidders had to be willing to bid, even though doing so risked revealing confidential information to competitors. Bidders reluctant to share their intentions would want to wait until near the end of the auction before bidding, as we saw in chapter 7 when we considered sniping in eBay auctions. But if *everyone* waited, the information needed to produce an efficient allocation wouldn't be transmitted.

To avoid this, the design for the spectrum auction included *activity rules,* proposed by my colleagues Paul Milgrom and Bob Wilson, to prevent bidders from making late bids unless they had made bids on equivalent numbers of licenses (measured in terms of population served) earlier in the auction. Thus big bidders had to make their bids known early, and all bidders could adjust their bids in light of the competition.

Simultaneous ascending auctions with activity rules enabled many bidders to compete simultaneously for many licenses, creating a thick market in which price discovery could take place. The activity rules also kept the auctions from dragging on interminably — another possible side effect of thick markets having to cope with the congestion of many possible transactions.

It wasn't long before these auctions were widely adopted by other countries interested in selling spectrum licenses. Today a number of European countries have gone further and implemented ascending auctions that allow *package bidding,* instead of requiring that the package be assembled license by license. A package bid is of the form, *I bid one hundred million dollars for precisely this package of licenses, and if I don't win the whole package, I don't want any part of it.* That is, a package bid allows a company to bid on exactly what it wants, and if it isn't the high bidder, it is completely free to bid on another package, without being constrained by its previous bids (as would be the case in a simultaneous ascending auction). That spares the company from becoming the high bidder for some licenses that it might no longer want.

Auctions with package bidding for many licenses can only be run now that a lot of computer power is available. In an auction for a single license, it's easy to tell what the winning bid is: it's the highest one. In a simultaneous auction for many licenses, it's still easy to tell what the winning bids are: they are the highest bids in the auctions for each license. But when many licenses are sold in a single auction that allows package bidding, it's a hard computational problem to determine which are the winning bids — that is, the bids for *packages* that would yield the highest value.

Suppose we were running a tiny auction for just four licenses, L1, L2, L3, and L4. One bidder might bid on the package consisting of L1 and L2, another might bid on L2 and L3, and a third might bid on L1 and L4. Notice that the first and second bidders can't both be winners, since they both want L2, but the second and third bidders can both get the entire package they want. So even if the first bidder makes the highest bid (for L1 and L2), the winning bidders might be the other two bidders, if the sum of their bids is greater than the first bid alone.

Of course, an ascending auction with package bidding has to determine the winning bidders at each stage of the auction so that bidders can be told whether their bids are currently winning the package they want, and losing bidders can formulate new bids. When there are lots of licenses for sale, lots of packages have to be considered. If there are even just four licenses for sale, there are already fifteen possible packages that could receive a bid (each of the four licenses separately, each of the six possible packages of two licenses, each of the four possible packages of three licenses, and the package consisting of all four licenses). To compute the set of winning bids at each stage of the auction, the auctioneer (i.e., his computer) has to take into account every combination of bids in which no two bids have a license in common in order to find the combination that yields the highest value.

Notice that when packages are being bid for, it may not even be possible to identify the price of each license, since the licenses are all bundled in packages of different sizes and compositions. So that's another way in which prices don't do all the work: as we saw in the example with just four licenses, the bidder who makes the highest bid may not even be among the winners. It's not only your bid that determines what you get but also the bids of others, not just on packages that compete with the one you want but also on packages that, together with yours, might add up to the highest value. So you can't just choose what you want, even if you're the richest bidder.

The very existence of our mobile devices is a result of the frequency auctions I've just described. And almost every time we use those devices, we drive another kind of auction that has become the financial lifeblood of the Internet.

Bidding for Eyeballs

The television show *Mad Men* depicts the advertising world of the 1960s. Back then, ads had to be targeted broadly, at a whole demographic, since the marketplace was run by newspaper and magazine publishers, broadcasters, and billboard companies. But on the Internet, ads aren't just targeted at people like you; they are targeted at people *just* like you, and often at *you* personally. That's because when you use the Internet, your eyeballs are for sale to the highest bidder, in some of the fastest auctions ever.

The business model that makes Google one of the most valuable companies in the world involves running auctions for words typed into its search engine. Every time you search, you not only see the "organic" results of the search for the words you have typed; you also see advertisements. The ads that you see, and the order in which they appear, depend on which advertisers win an auction conducted automatically by Google at the time of each search. Before the auction is conducted, advertisers have submitted bids based on the words you're searching for. Loosely speaking, the ad that appears in the first position on your screen was submitted by the highest bidder for your search words, who pays the price bid by the next-highest bidder, whose ad appears just below, and so forth, sometimes for a whole string of ads.

Online ads can be targeted to those who express a clear interest in a product, and the price of those ads can also be tailored to the value of the potential customer. (In the old days, by contrast, billboard space for a car ad cost the same as an ad for tea or soap.) For some years, the search word that drew the very highest bids was "mesothelioma," the name of the deadly disease that afflicts people who once worked with and inhaled asbestos. Because of the way the many liability cases involving asbestos were combined and settled, the law firms that handled such cases knew that someone who was searching for that disease might be a potential plaintiff who would quickly be entitled to a large settlement that his lawyers would share. So when someone searched for the disease, he saw ads for law firms.

You can see the online ad market at work by comparing the screens you see when you search for a noncommercial term, such as "mathematics," and for something that is widely sold, such as "new cars." The latter search will be full of ads, because someone searching for where to buy a new car is just the kind of person to whom auto dealers and manufacturers would like to direct their ads.

Needless to say, that Google auction has to be fast enough that you don't get tired of waiting and switch to a different search engine. A similarly speedy auction, which is also often targeted directly at you personally, takes place when you point your Web browser at a widely viewed Web page such as that of a newspaper. Just as when you look at a dead-tree version of the paper, you see ads on the page. But on the Internet, some of those ads have been placed there just as the page was loading onto your screen, by an ad exchange that auctioned the right to show an ad to you. Yes, to *you.*

The reason you can be targeted very precisely is that unless you regularly erase the "cookies" that websites put on your browser to track your viewing, ad exchanges—which auction the "banner ads" that appear on Web pages—can sometimes show advertisers precisely what you have recently been looking for.

For example, shortly after I moved to California, I decided to buy a treadmill desk. After a little Web browsing, I found one, and then I quickly began to see ads for treadmill desks every time I looked at the *New York Times* online. It's pretty expensive to buy an ad on the front page of the actual paper newspaper, and the only advertisers for whom that makes sense are ones who think it's valuable if a lot of people see their ads. But when my browser starts to load the electronic version of the newspaper and the cookies on my computer reveal that I might actually want to buy a treadmill desk, my eyeballs become very valuable property, worth bidding a high price for by someone selling an expensive item that only a few people want. That advertiser will bid a lot for my "prequalified" attention.

This may strike you as a little creepy. I was initially glad to see ads for treadmill desks, but they kept appearing even after I was already reading the news while pacing away at my new desk. As the treadmill-desk cookie on my browser got more and more out of date, I wondered if less and less was being bid for my eyeballs. And, of course, it made me very aware that many of the ads I saw were different from the ones other people would see even if they were reading the same story in the same newspaper at the same time. It felt like someone was watching me on the Web, which is in fact the case. My eyeballs, and yours, are for sale to the highest bidder.

If you want to preserve some privacy, you can take some steps yourself—such as erasing your cookies regularly. Or you may be glad to see ads that are tailored to what you may actually want, based on your searches, your Web browsing, and even your emails. But if you use the Web a lot, you would have to work pretty hard to stay completely private on your own. And it's not just your behavior online. The map software on your smartphone knows not only where you are but where you're going. Even your cell phone company has to know where you are whenever your phone is turned on, in order to be able to route calls to you, using the local radio spectrum license it bought at auction for that purpose.

So it might be, as all this new technology impinges on our privacy, that we'll want legal restrictions on some kinds of transactions involving our private data. Property rights—who owns what, and what they can do with it—are an important part of market design, and I predict that we'll be seeing some new efforts to define the property rights to our transaction data.

Property rights can be complex. We've already seen that you can donate a kidney to someone who needs a transplant, but it's against the law to buy or sell a kidney to be transplanted. So your kidney is your property in most ways—it's yours to keep or give away—but it isn't the kind of property you can sell. Let's look now at the surprising variety of other markets that are, in one way or another, *repugnant.*

PART IV

Forbidden Markets and Free Markets

11

......

Repugnant, Forbidden ... and Designed

IT'S ILLEGAL TO sell horsemeat at a restaurant in California. This isn't some leftover law from the Wild West, when a horse was a man's best friend. It's part of the California Penal Code that was enacted by popular referendum – that is, by direct vote – in 1998, long after horses had ceased to be an important part of the California economy. Section 598 of the penal code states in part: "Horsemeat may not be offered for sale for human consumption. No restaurant, cafe, or other public eating place may offer horsemeat for human consumption." The measure passed with 60 percent of the vote, with more than 4.6 million people voting for it.

This isn't a law that seeks to protect the safety of consumers by governing the slaughter, sale, preparation, and labeling of animals used for food. It's also different from laws prohibiting the inhumane treatment of animals, such as the rules on how farm animals can be raised or slaughtered, or the laws prohibiting cockfights.

In fact, it isn't illegal in California to kill horses; the California law only forbids such killing "if that person knows or should have known that any part of that horse will be used for human consumption." In other words, you can kill a horse in California and feed it to your dog; just don't eat it yourself. Ironically, the use of horsemeat in pet food has declined in the face of growing demand in Europe for U.S. horsemeat for human consumption.

Repugnant Transactions

Let's call a transaction *repugnant* if some people want to engage in it and other people don't want them to.

The kinds of repugnant transactions I'm most interested in are those in which it isn't easy to specify why some people object to them. Economists say transactions have "negative externalities" if they harm people who aren't party to the transactions. If your neighbor opens a nightclub and throws loud parties at 2:00 a.m. and the noise wakes you up, that's a negative externality. It's easy to understand why you object to those parties, even if everyone who attends them is a consenting adult who happily pays the cover charge and has a great time. Zoning regulations might well prohibit anyone from operating a nightclub where you live, precisely so that you can enjoy quiet nights.

Transactions with obvious negative externalities aren't my focus here, even though they may be transactions that some people want to engage in and other people don't want them to. So, for the purposes of this book, I'll reserve the word *repugnant* for transactions that some people want to engage in and that *are objected to by people who may not themselves experience any direct harm.*

Notice that repugnance is different from disgust. There's no law in California against eating worms or bugs. You can't find those meals in restaurants either, but that's because hardly anyone wants to tuck into a plate of fried worms. But California is a state with a population that hails from all over the world, including some places where horsemeat is considered delicious. Indeed, if you search for "boucherie chevaline" or "Pferdefleisch" on Google, you will be directed to gourmet horsemeat butchers in French- and German-speaking parts of the world.

So something can be repugnant in one place but quite all right in another, and repugnant to some people but not others. The reason it's against the law to eat horsemeat in California is not because no one wants to (in which case the law would serve no purpose), but because some people would like to and other people don't want them to. Of course, a transaction can also be repugnant but not illegal: in California before the 1998 law was enacted, many people found it repugnant that restaurants were allowed to serve horsemeat.

Sometimes transactions are legal even though repugnant because not enough people find them repugnant to turn that repugnance into law. Other times transactions are legal even though repugnant because it's too hard to enforce laws against things that enough people want to do, and trying to ban such transactions opens the door to black markets and crime. A classic case is the experience of the United States with banning the sale of alcohol.

In the name of public morality, the United States prohibited the sale of alcohol from 1920 to 1933 via the Eighteenth Amendment to the Constitution. This period, known as Prohibition, proved to be a disaster. It turned out that the national repugnance toward alcohol consumption and addiction wasn't nearly as deeply or broadly felt as first thought, and Americans quickly turned into a population of lawbreakers and bootleggers, in turn feeding organized crime. In the end, the Eighteenth Amendment was repealed by the Twenty-First Amendment, although a few individual states and counties still retain a variety of restrictions beyond the universal regulations that prohibit the sale of alcohol to minors or driving while intoxicated.

The repeal of Prohibition didn't just make alcohol legally available; it also knocked the underpinnings out from under the black markets that had provided such a thriving business for criminals.

But the criminal organizations that had grown rich under Prohibition moved into other businesses and were a long-lasting reminder of how prohibiting a market is a ham-handed form of market design that doesn't necessarily achieve even its main objectives.

It may help to explain repugnance to note that some transactions are just the opposite. Let's call a transaction *protected* if many people would like to promote it, in the sense that they are eager to protect others' rights to engage in it, even if they don't wish to engage in it themselves. Farming by small farmers falls into this category, since farming is subsidized around the world in ways that try to keep small farms viable in the face of encroachment by big (and efficient) agribusinesses.

Both repugnant and protected transactions are in the eye of the beholder. Religious worship is a protected transaction in the United States, enshrined in the First Amendment to the Constitution. But words such as *blasphemy, apostasy,* and *heresy* convey the repugnance that some people feel toward the ways other people worship. As I write this in 2014, wars are being waged between followers of different schools of Islam, much as Europe experienced wars over different strands of Christianity in centuries past. Similarly, the American right to own guns, protected by the Second Amendment, is nevertheless a hot political issue in the United States, in tension with gun control proposals related to the negative externalities that guns produce in many American communities.

So repugnance is different in different places, and it can last for a long time. But when it changes, it can change very fast.

One timely example is same-sex marriage. This is a transaction that some people want to engage in — they want to marry each other — and other people think they shouldn't. In most of the world for most of history, marriage, and the special social and legal status it offers as a protected transaction, was reserved for a man and a woman, or, in polygynous societies, a man and one or more women.

In the United States, same-sex marriage first became legal in a single state, Massachusetts, in 2004. The ban on gay marriage was lifted in Massachusetts by a court ruling that allowing only heterosexuals to get married violated the state constitution's guarantee of equal protection to all citizens. The Massachusetts court decision is an example of how a legally enforced repugnance was ended quite suddenly.

But same-sex marriage remains an issue about which Americans are divided. As I write in 2014, close to forty states have legalized same-sex marriage or are close to doing so (some through the courts, others through legislation), while a handful of others have actively reaffirmed their laws against it (although these bans may yet be struck down in court). Surveys suggest that repugnance toward same-sex marriage is now concentrated among older voters. So I suspect this repugnance will fade away with time.

When we look back at marriage, however, we can see that changes in repugnance over time go both ways. For example, polygamy, as chronicled in the biblical stories of King David and others, survived in various forms for many years and is still accepted in the Islamic world. But polygamy was forbidden among European Jews more than a thousand years ago, and it is illegal today in every American state. That said, there are dissident communities that openly practice polygamy in Utah and elsewhere, and of course private polygamy no doubt persists. Polygamy advocates have also begun to join the discussion involving the new same-sex marriage laws in order to question laws against plural marriage—suggesting that sometime in the future, history may turn again.

So it isn't the case that as we get all modern, we just abandon old repugnances. Sometimes we revisit them—or develop new ones.

Slavery is an obvious example of a market that is now repugnant and illegal even where it was once accepted, as it was in the United States. Of course, slavery wasn't a voluntary transaction as far as the slave was concerned, but today we find servitude so repugnant that a person may not even voluntarily sell *himself* into slavery or indentured servitude. Yet indentured servitude—term-limited slavery, entered into voluntarily—was once a common way for Europeans to buy passage across the Atlantic to America.

Today all forms of involuntary servitude are forbidden by the Thirteenth Amendment to the U.S. Constitution, ratified in 1865 after a bloody Civil War. It states: "Neither slavery nor involuntary servitude, except as a punishment for crime whereof the party shall have been duly convicted, shall exist within the United States, or any place subject to their jurisdiction."

Another important repugnance that has changed over time is lending money with interest. For centuries in medieval Europe, the church forbade charging interest on loans and enforced this ban on Christians. For a long time after that, the notion of charging interest continued to arouse repugnance. (Shakespeare devoted a whole play, *The Merchant of Venice,* to moneylending, and in *Hamlet* the Bard has Polonius advise Laertes, "Neither a borrower nor a lender be.")

That situation has obviously changed today, when the banking industry is a prominent part of the global economy (although Islamic law is commonly interpreted as forbidding interest as such). Finance is such a large industry (which can inspire a certain repugnance of its own) that it can be hard to fully appreciate what a giant change in public attitudes took place just a few centuries ago. But it's worth considering that change, to get some idea of the importance of public attitudes in determining what kinds of markets we allow.

Near the beginning of his long essay *The Protestant Ethic and the Spirit of Capitalism,* Max Weber quotes Benjamin Franklin on the virtues of responsible lending and borrowing. Franklin's view was the opposite of Polonius's: he felt that responsible borrowing and lending were Puritan virtues, and he offered advice about how to use credit responsibly. In 1748 he wrote an essay on the subject entitled "Advice to a Young Tradesman, Written by an Old One." (Franklin's essay is most famous for the aphorism "Remember that TIME is Money," but it also includes the parallel advice "Remember that CREDIT is Money.") Near the end of his own essay, Weber asks, "Now, how could activity, which was at best ethically tolerated, turn into a calling in the sense of Benjamin Franklin?"

Because markets are usually enmeshed in a web of connections to other markets, changes in repugnance can have far-reaching effects. For instance, evolving attitudes toward debt and involuntary servitude interacted with each other to change how we think about bankruptcy. In colonial America and the early years of the Republic, insolvent debtors could be imprisoned or sentenced to indentured servitude. But as involuntary servitude became more repugnant and debts less repugnant, bankruptcy laws were rewritten to be less punitive toward debtors.

The interconnectedness of markets sometimes allows their participants to avoid particular repugnant transactions but still achieve similar ends. For example, credit markets are very connected to asset markets, which is just to say that people borrow money in order to buy things. And although charging interest on borrowed money is repugnant under Islamic law, charging rent on assets is not. So while a conventional Western savings and loan company might lend you money to buy your house and charge you interest on the loan, in Islamic finance those transactions are sometimes structured so that a bank lends you money to buy a house, assumes part ownership of the house, and then charges you rent.

Just as repugnance toward existing transactions can change over time, new technologies make new kinds of transactions possible, and can in turn arouse new kinds of repugnance. Today it's feasible to purchase, in some places at least, the whole "supply chain" needed to produce a human birth. You can buy human sperm and eggs, and have the egg fertilized and then brought to term in a surrogate womb. This possibility has given rise to "fertility tourism" among people who are desperate to have children after ordinary methods have failed but who live in countries where surrogacy is illegal, or illegal to pay for. They find their way to countries where legal contracts can be made for these services. India is a big market for surrogate babies, and so, to a lesser extent, is the (more expensive) United States. But even in the United States, the laws vary, and as I write this in 2014, it is fully legal to pay a surrogate in California and many other states, but illegal in New York.

All these examples make clear that some kinds of transactions are repugnant in some places but not in others, and that repugnance can shift over time. Repugnance is very much in the eye of the beholder, and the beholder is observing other people's transactions. This makes repugnance hard to predict, let alone prescribe.

One common occurrence is that some transactions that may not be repugnant — and may even be protected — when no cash changes hands become repugnant when money is added to the mix. These are worth examining more closely, because they shed light both on repugnance itself and on the different kinds of markets and marketplaces that can, or sometimes can't, be designed to fill different needs.

Cash and Care: Buying Is Sometimes a Repugnant Way to Get Things

Some gifts and in-kind exchanges become repugnant once money is brought to the table.

The historical repugnance toward charging interest for loans seems to fall into this class, as do prohibitions on paying birth mothers of children put up for adoption and perhaps even on prostitution. Loans, adoption of children, and love are widely regarded as good things when offered freely, even when their commercial counterparts are regarded negatively.

We can all identify at least some occasions when we would agree that cash is inappropriate. For example, dinner guests at your home may reciprocate in kind by bringing wine or inviting you to dinner in return, but they would likely not be invited back if they offered to pay for their dinner.

Debates about what can be bought and sold for money touch on some of the most fundamental issues of democracy. We're pretty sure that votes shouldn't be bought outright, but we have considerable disagreement about what the role of money should be in political campaigns and decisions. During the Civil War, soldiers drafted into the Union army could pay substitutes to serve for them, but it was harder to avoid conscription for America's wars in the twentieth century.

At the end of the Vietnam War, the United States abolished conscription altogether and today has an all-volunteer force. Those volunteers are attracted to serve partly through improvements in the pay and benefits that come with military service, as well as by their sense of duty, patriotism, and adventure. Critics argued that this would result in the ranks of our armed forces being filled by low-income citizens who, practically speaking, had little choice, while wealthier citizens would be free to ignore their traditional duty to their country. Opinions may differ on how much this has come to pass, but the fears that it would be the very poorest of the poor who served have not been realized; not everyone who might like to join can qualify, and it's a badge of honor to serve. I'll come back to this in a moment.

To return to the question of kidney sales, virtually no one objects to kidney donation for transplantation. But many people clearly regard monetary compensation for organ donation as a very bad idea, maybe even the kind of bad idea that only bad people have.

Such concerns about the monetization of transactions seem to fall into three principal classes.

One concern is *objectification,* the fear that the act of putting a price on certain things—and then buying or selling them—might move them into a class of impersonal *objects* to which they should not belong. That is, they risk losing their *moral value.*

Another fear is *coercion,* that substantial monetary payments might prove coercive—"an offer you can't refuse"—and leave poor people open to *exploitation,* from which they deserve protection.

A more complex concern is that allowing things such as kidneys to be bought and paid for might start us on a *slippery slope* toward a less sympathetic society than we would like to live in. The concern, often not clearly articulated, is that monetizing certain transactions might not itself be objectionable but could ultimately cause other changes that we would regret. That slide might begin, for example, by reduced public enthusiasm for existing forms of support for the poor and vulnerable in ways that might make them feel that they *needed* to sell, say, a kidney.

When I speak to diverse audiences about these concerns in regard to kidneys, I find that lots of people nod their heads and think my words summarize why it's obvious that we shouldn't allow kidneys to be bought and sold. Another large group gets mad and thinks that people who don't want to sell kidneys shouldn't do so – but that they have no business stopping well-informed adults from engaging in an exchange that benefits both sides, that is voluntarily entered into, and that saves lives.

To help the two sides understand each other's viewpoints, I ask members of the audience to raise their hands if they're willing to consider carefully regulated sales of live kidneys. Even in a group of economists, not everyone raises his or her hand, and even in a group of noneconomists, some people do. I ask everyone to look around and get a sense of the distribution. Then I ask them, How about carefully regulated sales of live *hearts?* Selling hearts would, I remind them, kill the seller.

Most hands go down at that point, although there are almost always a few hardy souls who keep their hands up. It's not that there wouldn't be some supply, as well as considerable demand, for transplantable hearts: healthy people do sometimes kill themselves, and sometimes others as well, and they might be persuaded, instead, to also save a life and support their survivors financially if they could sell their heart. But most people still think that would be quite a bad idea.

My point is that most people find *some* transactions repugnant. That's a reason to treat other people's intuitions about repugnant transactions with respect, even if they don't raise and lower their hands at the same moments we do.

Repugnance as a Challenge to Market Design

What does treating repugnance with respect—even (or especially) other people's repugnance—mean for market design? In particular, how should we think about a situation in which there are some kinds of transactions that don't have the support needed to make a market, but for which there's substantial demand, and which might substantially improve some people's welfare?

Islamic finance is an example worth recalling. Islamic law forbids charging interest on loans, but people who live in Islamic countries, or who observe Islamic law wherever they live, still need to buy houses and other things, and they don't necessarily want to wait until they can pay cash. So they have the demand for something that would work like conventional mortgages or loans, even though they don't want to take out a loan that charges interest. A variety of financial instruments have been invented in response to this demand, which are more or less widely accepted as being compliant with Islamic law. They function somewhat like interest-bearing loans, but instead involve rent, deferred payments, or some other alternative way of structuring the transaction. These financial inventions bring to the Islamic world some of the big benefits that credit has brought to the wider world economy, and some of the hazards as well.

In a somewhat similar way, the widespread repugnance toward cash for kidneys, together with the equally universal shortage of kidneys, presents a big challenge for market design. The only country in which it is legal to buy and sell kidneys from living donor/sellers is the Islamic Republic of Iran. Legal markets were permitted there after the need for kidneys spiked during the Iran-Iraq War. Kidney donor/sellers in Iran also receive exemption from military service. We might be able to learn something of value about designing such markets by carefully studying the Iranian market. (Ironically, you could circumvent the repugnance toward interest on loans in Iran by financing a purchase with the proceeds from the sale of your kidney.)

Kidney exchange is a market design invention that succeeded in increasing the number of transplants through in-kind exchange of a kidney for a kidney, without running afoul of repugnance. Kidney exchange has not only become a standard form of transplantation in the United States, but it is spreading around the world. By U.S. law, a kidney must be a gift, either from a deceased donor or from a living donor.

But kidney exchange by itself doesn't end the shortage of kidneys. Today there are 100,000 people waiting for a kidney transplant in the United States alone. Meanwhile, we have only enough donated kidneys of any sort to perform about 17,000 transplants a year. That enormous shortfall—which wouldn't be filled even if every possible deceased donor donated two kidneys—is a heartbreaking reminder that altruistic donation alone is not filling the need.

Consequently, there's a lively ongoing debate about whether and how kidney donors might be treated more generously. Many doctors, hospitals, foundations, and patients argue that the law should be changed to allow living kidneys to be bought, so that the supply of kidneys can keep up with the demand.

Economists have long been accustomed to the fact that cash payments can fill such gaps by providing incentives to increase supply. Adam Smith, in his book *An Inquiry into the Nature and Causes of the Wealth of Nations* (1776), famously observed, "It is not from the benevolence of the butcher, the brewer, or the baker, that we expect our dinner, but from their regard to their own interest." Economists mostly think that allowing some incentives, monetary or otherwise, to be offered for giving someone a kidney could increase the supply of kidneys. In fact, there isn't much debate about that. So the repugnance toward kidney sales gives us the opportunity to understand repugnance and what it implies for markets in extremis, by looking at transactions that are forbidden even when lives would be saved by allowing them.

The debate, as well as the opposition to legalizing kidney sales, arises from concerns that this increase in supply, even if regulated as carefully as possible, would come with costs—to donor/sellers, to the poor and vulnerable, and to society at large—and that these costs might outweigh even the large benefits from saving many lives. Black markets, run by criminals, provide ample opportunity to see how those costs could be large. In at least some of these black markets, donor/sellers are deceived, coerced, not paid as promised, and almost never provided follow-up medical care. (The quality of the care provided to transplant recipients may also be suspect.) Moreover, black market kidneys are available only to the relatively rich and come from people who are quite poor.

There are also concerns that don't have to do with measurable costs. For example, a concern that has been expressed by the Catholic Church is that paying for kidneys inherently diminishes human dignity in a way that we should be reluctant to endorse even in a well-run market.

Market design can at least try to address the concerns about the costs that a market could impose on society. Legal markets are safer and easier to regulate than illegal ones. (Buying a bottle of wine today is very different from buying bootleg whiskey during Prohibition.) So it's worth thinking of market designs that might reduce or avoid the aspects of a market for kidneys that many find repugnant, in order to try to remove some of the barriers to transplantation that condemn many people around the world to early death and send others to illegal black markets, which thrive in many parts of the world.

Saving More Lives

Here are some preliminary proposals to think about, first about how a market in kidneys might be designed with cash compensation, and then, if that remains illegal, how kidney exchange could be extended to facilitate more transplants *without* cash compensation.

There has been a good deal of thinking about how some of the worst fears about compensating donors might be avoided with appropriate market design. For example, the concern that only the rich could afford kidneys might be addressed by amending the current outright ban on purchases to allow purchases only by a single authorized government buyer, with the obtained kidneys then being allocated according to the rules that today govern the allocation of deceased-donor organs.

The concern that sellers would be coerced by desperate circumstances could be addressed in part by having a one-year cooling-off period, in which prospective donors would be fully informed of the risks and benefits, as well as pass rigorous physical and mental health tests. And because a kidney transplant saves Medicare more than a quarter of a million dollars compared to continued dialysis, society could afford to pay generously enough that sellers who passed through those rigorous requirements wouldn't appear to be exploited.

Needless to say, there would be devils in the details. It may help to take a long-term perspective by considering how we would evaluate, twenty or thirty years after making it legal to pay for kidneys, whether legalization had been a good idea. Obviously, we'd want to determine whether the kidney shortage had been alleviated and whether the patients and donor/sellers were healthy and satisfied. We'd also want to know who the donors were and what had become of them. And I think we'd want to assess how both groups were regarded by the rest of society. This last issue is important because, once again, making a market legal doesn't necessarily remove the repugnance. (Prostitution is legal in Germany, but I'll bet no one running for political office there boasts of having been a sex worker.)

So what I would look for, twenty years after, is people campaigning for the U.S. Senate with the argument that you should vote for them because they cared enough to sell a kidney and save a life when they were young. That may sound far-fetched, but that's how Americans today regard the volunteer army. Soldiers are paid, but when they become candidates for political office, they boast about their military service and are honored for it. When we board an airplane at an American airport, uniformed members of the armed services are invited to board first. I'd be glad to line up behind kidney donors, too.

Another sign of success would be a long waiting list—not to get a transplant, but to sell a kidney—with self-help books authored by former donor/sellers bearing titles like *The Kidney Donor's Diet and Exercise Regime: You, Too, Can Qualify*.

Could this happen? Maybe, despite our best efforts, people who sold their kidneys would turn out to be poor, and ill, and exploited. It was presumably because of concerns like these that the Thirteenth Amendment made the market for indentured servitude illegal rather than trying to regulate it. And while I'm an optimist about what can be done through careful design and monitoring of a market to fix it if it isn't working well, I don't have a similar optimism about being able to successfully change laws that forbid kidney sales just about everywhere.

If cash markets for kidneys remain repugnant, market design still offers ways to expand the pool of possible donors without having to take on that repugnance directly.

For a start, we could learn about the role incentives might play in the decisions of organ donors by experimenting with removing *disincentives* for donation. Current American law does allow some money to change hands, in the form of paying donors for their housing, travel expenses, and lost wages. However, except in a few limited cases, most American donors bear these expenses on their own. (That's not true everywhere. In Israel, living kidney donors are now offered forty days' pay at their current wage, even if they don't miss that many days of work, and they are promised priority on the deceased-donor waiting list, in case they should ever need a transplant themselves.)

I would be glad to see some careful experiments, perhaps on a state-by-state basis, that would add evidence to the discussion of how donation would respond to payments. Even small payments might make some more donations possible. As with many diseases, kidney disease falls disproportionately on the poor, and so the potential donors for many patients are their spouses and close relatives and friends, who are poor themselves. But while even small payments may allow some donations to go forward, they would likely make only a small difference.

To take a big bite out of the big problem of kidney shortages, we'll have to do more.

Expanding Kidney Exchange

Kidney exchange is a good place to start, since it has been successful in increasing the number of donations and transplants without arousing repugnance. Kidney exchange is an in-kind exchange, kidneys for kidneys, a gift for a gift. And as I discussed in chapter 3, a big part of the success of kidney exchange has involved making good use of non-directed living donors to start chains of transplants. Some of those chains have been very long, and the average non-directed donor chain these days yields about five transplants.

But deceased donors are also non-directed, and they aren't being used to start chains. Not all deceased-donor kidneys would be attractive to patient-donor pairs hoping for a living-donor kidney, but *some* would be, as many deceased donors were young and healthy right up until a fatal accident.

Right now, each of the approximately 11,000 deceased-donor kidneys that become available each year in the United States produces one transplant. If we could instead include a substantial number of them in chains that would begin with the deceased-donor kidney going to an incompatible patient-donor pair and end with the living donor in such a pair donating a kidney to a person who was waiting for a deceased-donor kidney, we could get a lot more transplants.

Again, the devil would be in the details, because while everyone who is waiting for a kidney exchange is also on the deceased-donor waiting list, those individuals aren't at the very top of the list—that is, they haven't been waiting the longest. But if we could get the average deceased-donor kidney to facilitate just two transplants instead of only one, that would be a huge increase in transplantation, which would in turn shorten the wait for everyone.

Another possible solution would be to think about kidney exchange in a global way. There is virtually no kidney transplantation, and little or no access to dialysis, in places such as Nigeria, Bangladesh, and Vietnam, where kidney failure is a death sentence. Presumably, many kidney patients there have willing donors, but in a country such as Nigeria, for example, where fewer than 150 transplants occurred from 2000 to 2010, that willingness doesn't do patients any good. But suppose we were to offer them access to American hospitals, at no cost?

That may sound expensive, but it wouldn't have to be—indeed, it could be self-financing. Remember that removing an American patient from dialysis saves Medicare a quarter of a million dollars. That's more than enough to finance two kidney transplants, as well as postsurgical care and medicines. That money could pay for an exchange between an American patient-donor pair and, say, a Nigerian pair. We could push the envelope further (and perhaps risk arousing some repugnance, but be able to offer transplants to more foreign patients) if the foreign patients and donors would sometimes recruit a non-directed donor to accompany them. In this case, the living-donor surgery on the foreign pair could be financed by the savings on dialysis from a non-directed donation to a chain or to someone on the deceased-donor waiting list in the United States.

I can imagine conditions in which that kind of medical foreign aid would not only save foreign patients who would otherwise quickly perish but also cut the waiting time for kidney transplants in the United States to a fraction of its current, lethal length. It might also radically reduce the demand for illegal black markets. Talk about gains from trade.

Black and White?

I mention these proposals about kidney exchange partly to stimulate the important discussion of how we could alleviate the shortage of transplantable organs and increase access to kidney transplants. But I also want to use it as an example to emphasize the larger issue that even when thinking about the most difficult markets—those that may arouse our repugnance—we should never forget that markets are human artifacts. Market design allows us to think about how to try to get the benefits of markets to the people who need them.

For a repugnant market, "yes or no" isn't a black-and-white issue. Because markets are collective enterprises, we can design them but not necessarily control them. This is part of why there is so much sentiment in favor of making some markets illegal rather than trying to design them to circumvent their repugnant aspects. Markets unleash powerful forces, and so, the reasoning goes, if we can't completely control them, maybe we should ban them altogether when the risks seem high. The fact that laws banning various markets are so widespread means that we can't ignore repugnance as a constraint on markets.

Nevertheless, banning markets is just one way of trying to control them, and preventing markets is easier legislated than done. Making a market illegal stops legal markets. The markets we try to ban, repugnant markets, are precisely those that some people willingly take part in despite others' opposition. People wanting to transact with one another is a powerful force. The same force that has made markets an ancient and pervasive human activity also leads to black markets springing up where legal ones are prevented.

As the American experience with Prohibition shows, sometimes banning a market leads to widespread lawbreaking. Prohibition cut Americans' consumption of alcohol, but at great cost, and not nearly as much as it cut *legal* consumption. Something similar is going on today with efforts to ban not only narcotic drugs but also marijuana. We speak of a "war" on drugs, and indeed military weapons and assaults are often involved, sometimes against drug operations that control small countries and profoundly disrupt larger ones.

In the United States, drugs remain widely available, while our prisons are full of people swept up in the antidrug war. In California, where I live, marijuana is estimated to be among the top cash crops — in a state that serves an enormous agricultural market and grows more than 10 percent of the country's legal cash crops.

Some experiments with legalizing various aspects of drug consumption are slowly developing. Two states, Colorado and Washington, have legalized marijuana consumption and production even for recreational use, following a number of other states that had legalized it only for medical use. Other states no longer treat personal possession of marijuana as a criminal offense. European countries such as the Netherlands have had longer experience with the decriminalization of marijuana. And starting with Portugal in 1991, a number of countries have decriminalized the personal possession of all drugs.

To think about how we should judge whether relaxing the ban on a formerly banned market improves the situation or makes it worse, let's focus that question, hypothetically, on markets for narcotics such as crack cocaine. Let's further suppose that we agree that crack is an addictive drug that compromises the health of its users and has no medical use—that is, it's a drug that none of us wish to see used by anyone.

We already know that the war on drugs hasn't made crack unavailable, so we shouldn't compare the possibility of legalizing it only with the unattainable goal of eliminating it. We also don't have to consider legalizing it in an unregulated way—for instance, allowing it to be sold in schools. So the choices we need to consider aren't black-and-white.

Instead, we have to weigh the trade-offs. Legalizing the market for crack, even in a carefully regulated way, would very likely increase the number of addicts. That's a very bad thing. It might also decrease the amount of crime, and not merely the crime that would be eliminated by legalizing the possession and maybe the sale of crack, but also the crime that arises from having to consort with criminals to obtain it, and the crime that comes from having a commodity that generates an enormous flow of cash controlled by criminal methods to gain and keep market share. If the increase in addiction was huge and the decrease in crime was tiny, most of us would agree that we had made a bad situation worse. But if the increase in addiction was tiny and the decrease in crime was huge, I for one would think that we had made a good choice. Of course, even if we agree on how to evaluate the results of an experiment on legalizing crack cocaine, we may not agree at all on whether such an experiment would be desirable, or even ethical, depending on our possibly different beliefs about what the likely outcome of such an experiment would be.

To look at it another way, let's consider one other domain in which repugnant transactions are common: sex. People want to have sex with each other in circumstances that society disapproves of. But when we educate our children, pass laws, and try to control the transmission of disease, we would be foolish not to recognize that sex is a powerful force. That doesn't mean we shouldn't disapprove of some behaviors and try to moderate them, but sometimes we will more nearly achieve our goals if we try to channel behavior or offer alternatives, rather than institute a ban. (For this reason, we sometimes try to promote "safe sex" rather than abstinence.) In short, when we deal with sex, we need to recognize that we're dealing with powerful attractions.

Markets are like that, too.

12

Free Markets and Market Design

THINKING ABOUT THE design of markets gives us a new way of looking at them, noticing them, and understanding them. My hope is that this book will help you to see markets in new ways.

So may I take you to dinner to celebrate the completion of this book?

Field Guide to Restaurants

If we meet at my office at Stanford, we'll have a lot of choices about where to eat. Some nearby streets are lined with scores of different kinds of restaurants. We could go to University Avenue in nearby Palo Alto, or Castro Street in Mountain View, just a little farther away and the home of Google.

Both streets offer a *thick* marketplace for restaurant meals: not only are there lots of restaurants, but there are also lots of people who like to eat out. These restaurants chose their locations precisely because of all those customers, and despite the fact that they would be surrounded by competitors.

While all those restaurants are attracted to their thick marketplaces in similar ways, different kinds of restaurants deal differently with *congestion.* If we're going to eat at a popular time, we'll probably have to wait. And when I tell you how each restaurant deals with congestion—where and when we'll wait—you may be surprised at how many other things that detail will tell you about the restaurant.

Let's conduct a blind test: I'll tell you how three restaurants, call them A, B, and C, handle congestion on a busy night, and I'll bet you'll be able to figure out the color of their tablecloths.

If we want to eat in restaurant A, we call ahead for a reservation and then chat in my office until the time we are expected. (Or we can search the online marketplace OpenTable, which offers reservations to many restaurants, to compare the availability times of restaurant A and other, similar restaurants.)

When we arrive, we're seated quickly and given a menu. A server soon comes to ask if we'd like something to drink. When the drinks come, the server is ready to take our order, and we chat while the food is prepared. At the end of the meal, the bill is brought to our table, and after we've looked it over, we put down a credit card, which the server returns to take. When the server returns one last time with the credit card slip, we add a tip and sign it, and then get up and leave. Thus once we have arrived at restaurant A, almost all of the time we spend waiting is *after we are seated at our table.*

Restaurant B, by comparison, doesn't take reservations—and when we arrive there, the hostess takes our name and gives us an estimate of how long it will be before we can get a table. We can wait by the door (or outside if a lot of people are waiting), or we can wander off and return at about the appointed time. We can do this because, once we've given the hostess our name, we've reserved our place in line.

As soon as we're seated, a waitress comes to take our order and brings the meal quickly, along with any drinks we ordered. As we're finishing eating, she brings our bill. After leaving a tip in cash on the table, we take the bill to the cash register at the front of the restaurant, pay, and leave. Almost all of the time we spend waiting is *after we arrive at the restaurant but before we are seated.*

Finally, at restaurant C, we line up in front of one of the cash registers, quickly place our order with the cashier, pay, and are given our food on a tray. Then we find an empty table and eat. When we're done, we throw out our trash, deposit our tray in a stack to be washed, and leave. Most of what little time we spend waiting is *while we order and pay*, which we do in a single transaction.

What's going on here? Well, the three restaurants experience congestion at different parts of the dinner transaction and so deal with it in different ways. At restaurant A, the congestion is in the kitchen: the cooks prepare food to order, and the capacity of the kitchen prevents the restaurant from serving more customers quickly. We wait while they cook, and the pace of the restaurant is determined by the speed of the kitchen.

In contrast, at restaurant B, the food is mostly cooked ahead of time and just needs to be finished and plated. Here, the congestion is in the dining room: we have to wait for a free table, and the pace of the restaurant is determined by how quickly people eat.

Finally, restaurant C is a fast-food joint; the food is cooked assembly-line-style, so it is ready when we order it, and we only need to wait for someone to take our order.

So can you guess the color of the tablecloths? Well, I don't actually know the color at restaurant B, but there's a good chance the tablecloths are made of plastic and are wiped clean between customers. At restaurant A, the tablecloths are most likely white and are changed after each party leaves. There are no tablecloths at restaurant C.

No doubt, if you were to encounter a white linen tablecloth at a McDonald's or a tray and plastic tabletop at the Four Seasons, you would feel disoriented—and would worry that something had gone terribly wrong. And in terms of congestion, that would be true: the Four Seasons would gain nothing by being able to wipe down its tables quickly, while customers at McDonald's would be furious having to wait for the tablecloths to be stripped and replaced. Both establishments have found a way to deal efficiently with the kind of congestion they face.

Beyond Congestion

As we've learned, congestion is only one of the things that marketplaces must handle well. How about *safety*?

As you might imagine, safety in the market for restaurant meals has many dimensions. Can you be confident that you'll get the food and service that you expect, and that the food won't make you ill, and can the restaurant be confident that it will be paid? We've already seen in chapter 2 how credit cards helped restaurants deal with safety of payment. Now let's look at the other issues.

Since we are going to dinner near where I live, you can probably rely on my local knowledge of the quality of the food and service. I may have eaten in the restaurants before, and if not, I likely know people who have: local restaurants have a local reputation.

But even if you don't know a restaurant's local reputation, nowadays you can look at online crowd-sourced guides such as Yelp or Zagat, or (in the case of restaurant A) maybe the Michelin Guide or some other guide to fine dining. Restaurant C, on the opposite end of the dining spectrum, may be a franchise such as McDonald's, making it just a tiny node in the vast network of a corporation that strives to maintain comparable standards across all of its outlets. So if you're driving cross-country, you can pretty safely know what to expect even if you've never been to this particular restaurant before.

For all those restaurants, there's also another kind of safety provided by local government regulation. In this case, the County of Santa Clara Department of Environmental Health licenses restaurants and occasionally inspects them and issues reports. It can force a restaurant to close until any violations of the health code are fixed.

The health code is concerned with things that might often be difficult for customers to observe. Perusing the county's Food Facility Inspection Report reveals relatively few, mostly temporary closures of restaurants, for reasons such as "Food is subject to contamination from vermin" and "No paper towels and no soap at kitchen hand sink."

Other levels of government also play a role in regulating restaurants. For example, city zoning regulations prevent anyone from opening a restaurant in a residential neighborhood. And, as you now know, in California no restaurant can legally serve horsemeat.

Notice that McDonald's Corporation plays part of the same role with respect to a McDonald's franchise that the county plays with respect to restaurants in general: both have regulations to which restaurants must conform, and a restaurant that doesn't maintain certain standards can be forced to close.

Similarly, a shopping mall is owned by a private company that typically controls the precise number and variety of restaurants and other types of stores in the mall much more closely than cities are able to do through zoning. It isn't unusual for a mall to sign contracts with its tenants guaranteeing, for example, that a particular restaurant will be the only one in the mall that sells a certain cuisine.

Our field trip to eat dinner shows that details matter. Some restaurants have learned to use reservations to slow the flow of customers to cope with congestion in the kitchen, much as kidney exchange networks figured out how to organize nonsimultaneous chains to avoid operating room congestion. Mid-priced restaurants handle dining room congestion with waiting lines, and fast-food restaurants are fast not only because they prepare the food continuously, but also because they reduce each customer's transaction to a single encounter, at the cash register.

Like other marketplaces, restaurants have some common problems to solve, but the solutions to those problems may depend on the details of the markets and the kinds of transactions they deal in.

By the same token, the design of a marketplace can reflect some very local decisions, and some that are imposed on a range of markets by outside rule makers, public or private. These decisions may influence how *trustworthy* the market is, as well as how smoothly it runs.

Public and Private Regulation of Markets

Laws and regulations apply to a wide range of markets and marketplaces. Regulations can be supplied both by governments and by private entities, such as franchises, malls, or industry associations. But laws are the exclusive domain of governments.

Some laws provide the substructure for the design of many markets. Examples include laws concerning property rights (who owns what) and contracts (who agreed to do what in return for what else). The protection of property rights and the enforcement of private contracts depend in part on the courts, which are a government resource available not just to enforce laws but also to settle private contract disputes.

Property rights themselves need to be designed; they aren't all the same. Although you own your kidneys, and even have the right to give one away, you don't presently have the legal right to sell one anywhere except Iran. More familiarly, you may own the land on which your home is built, but local zoning laws may prevent you from selling food or opening a nightclub there.

If you bought a copy of this book, you own it and can choose to keep it, sell it, give it away, or review it. But you can't freely *copy* it. That's what copyright laws are for—they provide safety for authors and publishers. If you bought a *digital* copy of the book, your rights may be more restricted (you probably can't sell it) and are governed not by laws but by contract with the e-book publisher, similar to when you purchase software. When you buy an e-book or software, what you are in fact buying is a license to use it.

Like other elements of design, including rules made for individual markets, laws and regulations intended to govern multiple markets and marketplaces can have both good and bad effects. Health laws may protect us from unsanitary restaurant kitchens, but when they are slow to accommodate a changing market, they also "protect" us from gourmet food trucks. Similarly, McDonald's may allow its franchisees to introduce new products for health-conscious consumers more slowly than individual franchisees in California might like.

Good and Bad Design

While good market designs may emerge slowly over time as old rules and regulations are modified, bad designs can persist for a long time. To give an analogy from biological evolution, walking erect gives us humans many advantages, but we're not well designed for it; we suffer back problems and flat feet because we're made of parts that were largely shaped before we walked the way we do now. In the case of markets, bad designs can often persist not just because it takes time to discover better ones, but also because there may be lots of market participants with a stake in the status quo, and many interests are involved in coordinating any market-wide change.

That's one reason it isn't hard to find examples of markets that are not working well. The present system of health-care payments in the United States is a pastiche of poorly coordinated programs. Third-party payers—both private and public—that pass on the costs of care to subscribers or taxpayers don't necessarily have incentives to bring down costs, but neither do they always have incentives to exert every effort to ensure that people stay well. Under the present rules, it is in many ways harder to finance a clinic that educates diabetes patients on diet and other ways to control their disease than it is to finance the vastly more expensive dialysis and kidney transplants that become necessary when the disease progresses out of control.

Changing the market for health care is famously difficult; national political campaigns have been waged over it. More than four decades ago, President Richard Nixon tried and failed to set up a nationwide system that would provide health care for all. Only in the past couple of years has another attempt, the Affordable Care Act, or Obamacare, been enacted, and it is still hotly contested. But if I had to guess where the beginnings of good design might emerge, it would be in the health-care policies of large companies that self-insure their workers. Such companies benefit from keeping their workers healthy, as well as from reducing the costs of caring for them once they become ill.

Sitting here in California, another badly designed market whose consequences I see is that for water rights. As anyone who has seen the movie *Chinatown* or is familiar with the story of the Owens Valley knows, this problem is almost a century old.

Loosely speaking, water is allocated in the state without regard to how valuable it is in one place or another. For instance, we grow a lot of cotton in California, where it needs extensive irrigation. That irrigation water would be more valuable for other uses in California, particularly during a drought like the one we've been experiencing since 2012. But the people who own the rights to the water can't easily sell those rights when water is scarce, so they lose and we lose. Arranging a market for water rights shouldn't present barriers as high as those for making a cash market for kidneys, but so far the water isn't flowing where it's most needed.

Another market whose slow evolution perpetually surprises me is that for American residential real estate—that is, our homes. Professional real estate brokers, serving as very highly paid middlemen, continue to be almost unavoidable. They are paid a percentage of the selling price, often 5 percent, unlike everyone else who assists in the transaction—including the lawyers at the closing, who are paid by the hour.

Before the rise of the Internet, this may have helped make the market thick. It also may have dealt with congestion by helping buyers gather and filter information (and helping sellers signal it), since it was hard in those days to get information about houses. But nowadays many buyers take a virtual tour before deciding which houses to visit, and so much of the market doesn't really need a middleman.

Standard contracts in the industry, however, make it difficult to avoid paying for a broker's services even if you make little use of them. If the seller of a house has engaged a broker, the contract says that if the buyer also comes with a broker, the two brokers will split the fee – but the full fee must be paid even if the buyer comes without a broker. So once a seller has engaged a broker, she can't save on her brokerage fee even if the buyer finds the house without assistance.

One incremental, entrepreneurial market response is that there are now a few licensed brokers who seek to serve buyers who don't really need a broker. When my wife and I bought a house that we found without a broker, the seller had already signed a contract with a broker that fixed the total brokerage fee. When it came time to close the deal, we engaged a broker from the firm Redfin, which refunds half of its share of the fee to the buyer. That is far from an efficient adaptation to a changing world, but it's a start.

Computerized Markets

Although the Internet has so far failed to transform the housing market, computerized markets have, as we have seen, made enormous changes in other areas. Computers not only make markets ubiquitous and fast; they also make it possible to operate "smart markets" that depend on computational power.

Neither kidney exchanges nor "package bidding" auctions would be possible if computers weren't there to handle the gnarly calculations needed to find the best way to match lots of patient-donor pairs with one another, or to find the set of packages of spectrum licenses that will raise the most revenue at each point in the bidding. And computational speed is much more than just a convenience; it's another way that computers make new markets possible. Google couldn't auction advertisements based on search words if the auction for those words had to be conducted by a human auctioneer.

Free Markets

How do we square market design with the notion of the "free market" that so many people hold dear?

In chapter 1, I made the analogy between a free market with effective rules and a wheel that can rotate freely because it has an axle and well-oiled bearings. I could have been paraphrasing the iconic free-market economist Friedrich Hayek, who in his 1944 free-market manifesto *The Road to Serfdom* wrote, "There is, in particular, all the difference between deliberately creating a system within which competition will work as beneficially as possible and passively accepting institutions as they are." He understood that markets need effective rules in order to work freely.

Hayek also understood that there is a place for economists to help in understanding how to design markets. Using the word *liberal* in a slightly different way than we do today (he meant something closer to what we might call *libertarian*), Hayek wrote, "The attitude of the liberal towards society is like that of the gardener who tends a plant and, in order to create the conditions most favorable to its growth, must know as much as possible about its structure and the way it functions."

Coming back to current word usage, liberals and conservatives often find themselves in disagreement about the proper scope of government regulation of markets. Debates about markets often use the phrase "free markets" as a slogan, sometimes as if markets work best without any rules other than property rights. Hayek had something to say about that, too: "Probably nothing has done so much harm to the liberal cause as the wooden insistence of some liberals on certain rules of thumb, above all the principle of laissez faire."

As in a garden, only some plants grow without any help at all, and some of those are weeds.

The lesson of market design for political debate is that to understand how markets should be operated and governed, we need to understand what rules particular markets need. That's a different question than whether some rules will apply to many markets, as regulations, and whether government is best suited to make some of those rules.

The fact is that both governments and private market makers have a role to play, and they both can sometimes err by regulating too slowly and not vigorously enough, but also by regulating too hastily. (Happy is the nation that doesn't make steam engines illegal as soon as they're invented, when it looks like they might sometimes explode or put honest laborers out of work, but before it becomes clear how they will lead to an industrial revolution.)

When we think about helping markets work well, we also need to think about what it means to work well. Markets that function well give us choices, and so markets that operate freely are related to our liberty as well as to our prosperity.

We've seen how our choices may be limited in markets that aren't thick, or are congested, or that make it unnecessarily risky to even try to get what we would choose if we could. And, of course, markets are connected: some people may come to the marketplace without what they need to be offered good choices—for example, without having had an opportunity to go to a good school. That's why increasing school choice may increase choices in other markets as well.

Good designs are a moving target. Some markets may suffer because of rules that have yet to be made, while others may suffer from rules that have yet to be changed.

To use an analogy from civil engineering, the Romans built great roads and bridges, yet we don't build them the same way today.

That's partly because we have new materials and techniques and understandings that let us build bridges longer and stronger. But it's also because bridges, like markets, change the way people behave. Better bridges encourage more traffic, which creates congestion and which requires better roads and eventually bigger bridges. And those new roads and bridges mostly have to be built so as to connect to the already existing network of roads and bridges.

Market design isn't static either, and it also often proceeds by making incremental changes that allow it to connect with existing practices and with other markets.

The Language of the Marketplace

We encounter markets through marketplaces, just as we experience language through speeches, conversations, books, essays, and tweets.

And markets are like languages. Both are ancient human inventions. Both are tools we use to organize ourselves, to cooperate and coordinate and compete with one another, and ultimately to figure out who gets what. These two fundamental human artifacts play a role in all the things we do and in everything we make (we can't even make love, let alone war, without them).

Markets and languages both constantly adapt. There are a lot of words in modern languages that didn't exist when Sumerian was the language of commerce, and you can search for and buy many of the new things named by those new words on Amazon, using your smartphone. And there are specialized markets, such as kidney exchange, that are custom-designed to accomplish what more conventional markets cannot—just as there are specialized mathematical and computer languages to communicate things that elude ordinary speech.

Markets, like languages, come in many varieties. Commodity markets are impersonal, but matching markets can be deeply personal, as personal as a job offer or a marriage proposal. And once you observe that matching is one of the major things that markets do, you realize that matching markets—markets in which prices don't do all the work, and in which you care about whom you deal with—are everywhere, and at many of the most important junctures of our lives.

When we learn to listen and speak, as well as to read and write, we learn rules of courtesy, grammar rules, and shared vocabularies that have evolved even though no one planned them. In much the same way, markets and marketplaces, whether consciously designed (such as Amazon or kidney exchange) or developed incrementally by accident and happenstance, have rules that help them work well . . . or badly.

Economists as Engineers

So the design of markets, via marketplaces, is an ancient human activity, older than agriculture. And yet, after more than 10,000 years, it is still not widely (or deeply) understood. Economists used to study markets as if they were natural phenomena, much like the way we think of languages. They were seen as not being really under our control.

But the reason natural languages aren't really under our control is because they emerge from the interaction of millions of users. English speakers certainly know that it's hard to change the spelling of a word that could be spelled more simply and in a way that is closer to how it is pronounced. (Ev'n tho difrent spelings mite make sens, it's hard to get enuf peepul to agree.) Of course, the situation is different for artificial languages, such as those for computer programming. Indeed, as we've seen, the operating systems of computers and smartphones are themselves marketplaces of a sort. What makes those artificial languages—and other marketplaces—different from natural languages is that they have proprietors and groups of influential users who can coordinate to make needed design changes.

As we start to understand better how markets and marketplaces work, we realize that we *can* intervene in them, redesign them, fix them when they're broken, and start new ones where they will be useful. The growing ability in recent years of economists to be engineers is a bit like the epochal transformations that farming or medicine have experienced over the millennia.

The first farmers grew what they found, but over time farmers began to keep the seeds from their most successful crops to plant the following year, and so they started, inadvertently, to be plant breeders. Today we benefit from centuries of deliberate plant breeding, and more recently from genetic engineering to modify crops to have higher yields in more challenging environments. When we take a walk in a field or garden today, we see the result of generations of cultivation. Some of the plants we see are ancient species that remain well adapted to modern conditions, and others are modern cultivars, variations bred to thrive better, be more nutritious, or perhaps just to look prettier than their ancestors. Even the newest plants may still be pollinated by bees: they are part of an ecology that reflects the complex interaction of evolution, coevolution, and human desires and designs.

Medicine has made similar progress. Not so very long ago, doctors mostly specialized in telling you what was going to happen to you and in keeping you comfortable while it happened. Today we expect our doctors to intervene in many of our diseases, and they have drugs and surgical techniques that often enable them to succeed. We anticipate that medicine will be able to do even better in the future, but we're glad to benefit from what it can do now.

Because markets and languages are tools that we use collectively, they may be hard to redesign even when they are working badly. As a result, we have to limp along with some bad designs, like awkward spellings.

But sometimes we do get to redesign markets that are working badly. And sometimes we even get to design entirely new markets. These are opportunities to treasure, to study, to approach with humility, and to monitor with care.

Markets are human artifacts, not natural phenomena. Market design gives us a chance to maintain and improve some of humanity's most ancient, essential inventions.

Notes

1. INTRODUCTION: EVERY MARKET TELLS A STORY

page

14 *who gets what—and why:* As I said in my Nobel lecture, understanding who gets what, and how and why, is still very much a work in progress. You can watch the lecture online at http://www.nobelprize.org/nobel_prizes/economic-sciences/laureates/2012/roth-lecture.html.

2. MARKETS FOR BREAKFAST AND THROUGH THE DAY

15 *sold "by sample":* See Jonathan Levin and Paul Milgrom, "Online Advertising: Heterogeneity and Conflation in Market Design," *American Economic Review* 100, no. 2 (May 2010): 603–7.

23 *Credit cards offered merchants:* Credit cards are sometimes referred to by economists as "two-sided markets" because of the way they form a marketplace that needs to attract two different kinds of participants: merchants and consumers. One important strand of work focuses on how the two sides of the service should be priced; see, for example, Jean-Charles Rochet and Jean Tirole, "Two-Sided Markets: A Progress Report," *RAND Journal of Economics* 37, no. 3 (Autumn 2006): 645–67.

25 *they seldom switch cards:* See Lawrence M. Ausubel, "The Failure of Competition in the Credit Card Market," *American Economic Review* 81, no. 1 (March 1991): 50–81.

26 *middlemen:* For competition among middlemen, see Benjamin Edelman and Julian Wright, "Price Coherence and Adverse Intermediation" (working paper, Harvard Business School, Cambridge, MA, December 2013).

3. LIFESAVING EXCHANGES

32 *Lloyd Shapley and Herb Scarf:* Lloyd Shapley and Herbert Scarf, "On Cores and Indivisibility," *Journal of Mathematical Economics* 1, no. 1 (March 1974): 23–37.

34 *safe for them:* A. E. Roth, "Incentive Compatibility in a Market with Indivisible Goods," *Economics Letters* 9, no. 2 (1982): 127–32.

36 *We posted our paper:* The Web version was posted as NBER (National Bureau of Economic Research) Working Paper No. w10002 in September 2003, and the published paper came out in 2004: Alvin E. Roth, Tayfun Sönmez, and M. Utku Ünver, "Kidney Exchange," *Quarterly Journal of Economics* 119, no. 2 (May 2004): 457–88. You can check it out online at http://web.stanford.edu/~alroth/papers/kidney.qje.pdf.

38 *In 2005, we wrote:* One of the funny things about publishing papers in both economics and medicine is that economics papers take much longer to appear. That 2005 paper eventually appeared in 2007 as Alvin E. Roth, Tayfun Sönmez, and M. Utku Ünver, "Efficient Kidney Exchange: Coincidence of Wants in Markets with Compatibility-Based Preferences," *American Economic Review* 97, no. 3 (June 2007): 828–51. In the meantime, a follow-up paper that reported on a three-way exchange had already appeared in 2006 as Susan L. Saidman, Alvin E. Roth, Tayfun Sönmez, M. Utku Ünver, and Francis L. Delmonico, "Increasing the Opportunity of Live Kidney Donation by Matching for Two- and Three-Way Exchanges," *Transplantation* 81, no. 5 (March 15, 2006): 773–82.

one actual three-way exchange: While we were trying to make it easier for other kidney exchange programs to try three-way swaps like this one, they generated some controversy. In 2005, a group of doctors from Johns Hopkins Hospital in Baltimore published a paper in the *Journal of the American Medical Association* proposing a two-way kidney exchange algorithm that looked a lot like the one we had proposed, but that ignored the elements of market design that made it safe for patients and surgeons to participate. They proposed that a national exchange should be limited to two-way transactions, even though our recent work had already demonstrated the benefits of larger exchanges.

43 *In a 2006 paper:* Alvin E. Roth, Tayfun Sönmez, M. Utku Ünver, Francis L. Delmonico, and Susan L. Saidman, "Utilizing List Exchange and Undirected Good Samaritan Donation Through 'Chain' Paired Kidney Donations," *American Journal of Transplantation* 6, no. 11 (November 2006): 2694–2705.

45 *nonsimultaneous chains:* The title of our *New England Journal of Medicine*

article reporting that first chain was "A Nonsimultaneous, Extended, Altruistic-Donor Chain"—a NEAD chain for short. Mike had wanted a more flamboyant name with the same acronym: "Never-Ending Altruistic Donor chain." Heleena McKinney makes me think that Mike may have been on to something. That paper had a large and varied set of authors, including surgeons, economists, and computer scientists: Michael A. Rees, Jonathan E. Kopke, Ronald P. Pelletier, Dorry L. Segev, Matthew E. Rutter, Alfredo J. Fabrega, Jeffrey Rogers, Oleh G. Pankewycz, Janet Hiller, Alvin E. Roth, Tuomas Sandholm, Utku Ünver, and Robert A. Montgomery, "A Nonsimultaneous, Extended, Altruistic-Donor Chain," *New England Journal of Medicine* 360, no. 11 (March 12, 2009): 1096–1101.

48 *frequent-flier program:* Itai Ashlagi and Alvin E. Roth, "Free Riding and Participation in Large Scale, Multi-hospital Kidney Exchange," *Theoretical Economics* 9 (2014): 817–63.

49 *long nonsimultaneous chains:* Itai Ashlagi has taken the lead in understanding why long chains became important and how to manage them. See, for example, Itai Ashlagi, Duncan S. Gilchrist, Alvin E. Roth, and Michael A. Rees, "Nonsimultaneous Chains and Dominos in Kidney Paired Donation—Revisited," *American Journal of Transplantation* 11, no. 5 (May 2011): 984–94, and Itai Ashlagi, Duncan S. Gilchrist, Alvin E. Roth, and Michael A. Rees, "NEAD Chains in Transplantation," letter to the editor, *American Journal of Transplantation* 11, no. 12 (December 2011): 2780–81.

hard-to-match pairs: The hard-to-match pairs are mostly those with highly sensitized patients—that is, patients who have antibodies that make it hard for them to accept a kidney from almost anyone. The test that determines how sensitized a patient is involves a nice story. It was invented by the UCLA medical scientist Paul Terasaki, who also built a prosperous business making those tests available. His is the story of a remarkable American life and career. Born in California in 1929, he and his family were interned with other Japanese Americans during World War II. In 2010, he donated $50 million to UCLA. And in 2012, Itai Ashlagi and I shared the Terasaki Medical Innovation Award given by the NKR for our work on kidney exchange. In particular, Itai developed algorithms and software, now widely shared, that allowed the NKR to consider how best to combine chains of various lengths—including very long chains—with exchanges in short cycles to produce the most transplants over the long term.

50 *a very long way to go:* In the meantime, the potential computational challenges facing a national exchange, if not as thorny as the political ones, are still potentially formidable. The software that Utku had originally written to implement our matching algorithm for NEPKE and the APD could handle only up to about 900 pairs. While no one had ever assembled that many candidates for kidney exchange (and no one has yet), we were hopeful that a national exchange might eventually deal with many more patient-donor pairs than

that. One of the computer scientists who took up the challenge was Tuomas Sandholm, at Carnegie Mellon University in Pittsburgh. A graduate student of his, David Abraham, took a market design course that Utku was teaching at the University of Pittsburgh. Together with a third computer scientist, Avrim Blum, they figured out how to perform the kind of matching we had proposed for up to 10,000 pairs, more than we would expect in the foreseeable future. In fact, as we have started to gain experience with kidney exchange, it appears that we might be able to eventually stabilize the size of the pool at some fairly low level, as new transplants will balance out new enrollments.

51 *how to reimburse the costs:* For some thoughts on how reimbursements might be designed, see Michael A. Rees, Mark A. Schnitzler, Edward Zavala, James A. Cutler, Alvin E. Roth, F. Dennis Irwin, Stephen W. Crawford, and Alan B. Leichtman, "Call to Develop a Standard Acquisition Charge Model for Kidney Paired Donation," *American Journal of Transplantation* 12, no. 6 (June 2012): 1392–97.

Thousands of transplants: I occasionally get emails from kidney patients seeking advice about transplantation. Often they are seeking a donor. I don't have much specific help to offer them, but perhaps the generic information I send to them will be useful to others. This advice is for a kidney patient who is already registered with an American hospital that does a lot of kidney transplants.

If you are not already registered on the deceased-donor waiting list, talk to your doctors about getting on the list, since the amount of time you have been on the list plays an important role in assigning kidneys. Be aware that the waiting list is organized by region, and the wait can be much longer in some regions of the country than in others. (That's why Apple CEO Steve Jobs, who lived in California, got a liver transplant in Tennessee.)

A new organization called OrganJet helps people register on the waiting lists of regions where the wait is shorter. This group is mostly involved in helping arrange transportation (since patients have to be able to travel to the distant hospitals with which they are registered for checkups and other reasons). But its website says it's also equipped to offer advice on how to register with a transplant center in one of the regions with a shorter waiting time. That might be a good place to start, since there may be a conflict of interest between you and your local transplant center, which might not be interested in having you register at another transplant center.

A living donor, if you can find one, may be a better and quicker alternative. To begin, you might want to contact the Living Kidney Donor Network founded by Harvey Mysel. There are also various kidney matchmaking sites, such as MatchingDonors.com, and other, more specialized sites, such as KidneyMitzvah.com and Renewal (http://www.life-renewal.org/). My impression is that quite a few donors come from faith-based organizations, so if you are a member of a religious congregation, you might let the other members know about your search for a donor.

If you are considering a living donor, kidney exchange means that the donor you find needn't be compatible with you; he or she simply needs to be healthy enough to donate a kidney and willing to donate one, so that you can get a kidney in return. One of the several kidney exchange networks can take it from there. It is probably best to work with the one with which your transplant center has the best working relationship.

4. TOO SOON

57 *sorority rush:* See S. Mongell and A. E. Roth, "Sorority Rush as a Two-Sided Matching Mechanism," *American Economic Review* 81 (June 1991): 441–64.

58 *that was the plan:* See Michael Malone, *Charlie's Place: The Saga of an American Frontier Homestead* (Palisades, NY: History Publishing, 2012), 32–33.

59 Jumping the gun: For an account of many unraveled markets, see A. E. Roth and X. Xing, "Jumping the Gun: Imperfections and Institutions Related to the Timing of Market Transactions," *American Economic Review* 84 (September 1994): 992–1044.

62 *abandoned the attempt:* John Swofford, chair of the NCAA's Postseason Football Subcommittee (and athletic director at the University of North Carolina), explained this decision in a letter to me dated March 15, 1991: "The decision was made to eliminate from the NCAA Bylaws legislation that prohibits an institution from tying into a bowl before a particular date because that particular piece of legislation was being largely ignored and, most importantly, could not be enforced. In recent years, the NCAA has worked toward eliminating rules that were unenforceable, and the membership overwhelmingly felt that this was one of those rules. The bowl association has on its own decided to implement its controls, and there will continue to be a selection date, although it will not be an NCAA violation if an institution does not adhere to that date . . . Whether or not this will improve the situation, of course, remains to be seen. If this does not work, our committee is looking at the possibility of instituting a draft whereby teams would be ranked and given a drafting order, and the teams would be allowed to pick the bowl they would like to attend, or the bowls would be ranked in a particular drafting order and they would be allowed to draft teams into their bowl. Either of these would take place on a pre-determined date."

64 *Nielsen ratings:* For a detailed description of how the market for college football games changed from year to year, see Guillaume Fréchette, Alvin E. Roth, and M. Utku Ünver, "Unraveling Yields Inefficient Matchings: Evidence from Post-Season College Football Bowls," *RAND Journal of Economics* 38, no. 4 (Winter 2007): 967–82.

71 *"If you are 20":* Yael Branovsky, "Barely 16 and Married," *Israel News,* September 26, 2010, http://www.ynetnews.com/articles/0,7340,L-3959289,00.html. Roth and Xing, "Jumping the Gun."

72 *twice as many men:* Today there are more women than men in college. For the history of this reversal, see Claudia Goldin, Lawrence F. Katz, and Ilyana Kuziemko, "The Homecoming of American College Women: The Reversal of the College Gender Gap," *Journal of Economic Perspectives* 20, no. 4 (Fall 2006): 133–56.

73 *more widespread phenomenon:* Even the market for books like this one is unraveled, with publishers often buying manuscripts — and authors selling them — well before they are fully written, leaving both parties to work out the eventual quality of the match — and the book.

77 *Exploding offers aren't a problem:* Muriel and I were lucky to work with Dr. Debbie Proctor, a Yale gastroenterologist who became the motive force in reforming the gastroenterology fellowship market. For an account of the clearinghouse's early success, see Muriel Niederle, Deborah D. Proctor, and Alvin E. Roth, "The Gastroenterology Fellowship Match — The First Two Years," *Gastroenterology* 135, no. 2 (August 2008): 344–46.

5. TOO FAST: THE GREED FOR SPEED

82 *trades are made:* To be clear, the continuous electronic limit order book works by offering at any time the best bid and ask being offered by anyone on the market. The best bid is the highest price at which there is currently an offer to buy, so it's the highest price at which you can sell if you want to sell immediately. The best ask is the lowest price at which anyone is willing to sell, so it's the lowest price at which you can buy immediately. You can also make your own bid or ask, and if it is the best one, it becomes the market bid or ask; if it is not, it is queued up behind the better ones, and as other trades are executed, or as bids and asks are withdrawn, yours may become the best.

85 *well-established laws:* In a related matter, the news service Reuters used to share the results of a survey of consumer sentiment with some of its favored customers 2 seconds before it released the survey to the public on its newswire. Two seconds may not seem like much, but in the 10 milliseconds just before the public release of the news, hundreds of thousands of shares of S&P 500 exchange-traded funds would be sold.

That worried New York State attorney general Eric Schneiderman, who started referring to it as "Insider Trading 2.0." In the summer of 2013, he reached an agreement with Reuters not to release the survey results to anyone before reporting them to the public. Those 2 seconds made a difference: the next time Reuters released the survey, only five hundred shares were traded in the 10 milliseconds just before the results were shared on the newswire. That was a normal number for such a thick market. The headline of the related *New York Times* story published the next morning was FAIR PLAY MEASURED IN SLIVERS OF A SECOND.

86 *markets run only once:* See Eric B. Budish, Peter Cramton, and John J. Shim, "The High-Frequency Trading Arms Race: Frequent Batch Auctions as a Market Design Response" (working paper, Booth School of Business, University of Chicago, December 2013).

88 *a better design:* There may, of course, be other changes in the current design of the markets that could help competition by price regain ascendancy over competition by speed, and some of those changes might be easier to get adopted. For example, as I write this in 2014, two Stanford graduate students, Markus Baldauf and Joshua Mollner, are following up on the work of Budish and his colleagues by proposing a new design based on inserting some delays in how quickly offers may be accepted compared to how quickly they may be canceled, to protect liquidity providers in a different way from having their stale bids and asks sniped.

90 *a better market:* See Claudia Steinwender, "Information Frictions and the Law of One Price: 'When the States and the Kingdom Became United'" (working paper, London School of Economics and Political Science, October 2014).

I received the offer: Christopher Avery, Christine Jolls, Richard A. Posner, and Alvin E. Roth, "The New Market for Federal Judicial Law Clerks," *University of Chicago Law Review* 74 (Spring 2007): 448.

93 *courtroom gladiators:* Alex Kozinski, "Confessions of a Bad Apple," *Yale Law Journal* 100 (April 1991): 1707.

96 *"Increasing numbers":* Stanford Law School, "Open Letter to Federal Judges About Clerkships from Dean Larry Kramer," *SLS News,* July 17, 2012, http://blogs.law.stanford.edu/newsfeed/2012/07/17/open-letter-to-federal-judges-about-clerkships-from-dean-larry-kramer/.

"Although the judges": Judge John D. Bates to All United States Judges, memorandum, January 13, 2014, Administrative Office of the United States Courts, Washington, DC, https://oscar.uscourts.gov/assets/Federal_Law_Clerk_Hiring-January_13_2014.pdf.

97 *Hopwood:* Hopwood was hired by Rogers before beginning his third year of law school at the University of Washington. But he didn't get to law school the way most students do. After dropping out of college, he started a brief career as a bank robber, which ended with a sentence of twelve years in prison. While in prison, he became an accomplished jailhouse lawyer who helped other prisoners prepare successful petitions to the U.S. Supreme Court. After serving his time, he married and had children, and was later admitted to law school.

To practice law, Hopwood will eventually have to be admitted to the bar association of the state in which he practices. Bar associations impose character and fitness considerations, and unpardoned felons are typically excluded from becoming lawyers. I predict and hope that Hopwood will be an exception, and will become the rare lawyer who has experienced the law from both the very bottom and the very top. Note that judges are able to make their own rules about which clerks they hire, as well as about how they hire them.

The history of this market: The similar histories of unraveling in the market for law clerks (in which salaries are set by Congress) and the market for new lawyers at big law firms (in which salaries are set by competition) also make it clear that competition by speed can go hand in hand with price competition. See, for example, Alvin E. Roth, "Marketplace Institutions Related to the Timing of Transactions: Reply to Priest (2010)," *Journal of Labor Economics* 30, no. 2 (April 2012): 479–94.

98 *a series of agreements:* Among the parties involved were the Japan Federation of Employers' Associations (Nikkeiren), the Association of National Universities, the Ministry of Education, and, later, the Ministry of Labor.

informal guarantees of employment: Rules that prohibited firms from making formal offers of employment before a certain date were circumvented through informal guarantees of employment, known as *naitei.* As a result, employment decisions for university graduates unraveled. The popular name for this unraveling was *aota-gai*, which translates literally as "harvesting rice while it is still green."

even earlier ones: The newspaper *Asahi Shimbun* reported in 1970 that *aota-gai* was being replaced by *sanae-gai*—"harvesting rice while it is newly planted." Large banks began holding their employment exams months earlier than the rules allowed, and there were cases of *naitei* to students more than a year before they expected to graduate.

6. CONGESTION: WHY THICKER NEEDS TO BE QUICKER

103 *time-wasting false leads:* Airbnb took other steps to eliminate false leads. For instance, once you request a reservation, that spot on the host's calendar is marked as unavailable, although the host is still free to decline your request.

110 *"Before you might":* David M. Herszenhorn, "Council Members See Flaws in School-Admissions Plan," *New York Times,* November 19, 2004, http://www.nytimes.com/2004/11/19/education/19admit.html.

7. TOO RISKY: TRUST, SAFETY, AND SIMPLICITY

117 *eBay feedback system:* Recognizing the importance of trustworthy payments, eBay later bought PayPal, which eventually grew to be almost equal in revenue to eBay's main business. At that point, eBay made plans to spin off PayPal and again become two companies, which is happening as I write this in 2014.

118 *a new feedback system:* You can read about the careful market design and testing that went into eBay's decision to redesign its feedback system in Gary Bolton, Ben Greiner, and Axel Ockenfels, "Engineering Trust: Reciprocity in the Production of Reputation Information," *Management Science* 59, no. 2 (February 2013): 265–85.

more detailed and useful: Airbnb's feedback system has undergone a similarly motivated change, with neither party now able to see the other's review beforehand, to reduce reciprocal feedback.

120 *eBay snipers:* For more about sniping on eBay, see Alvin E. Roth and Axel Ockenfels, "Last-Minute Bidding and the Rules for Ending Second-Price Auctions: Evidence from eBay and Amazon Auctions on the Internet," *American Economic Review* 92, no. 4 (September 2002): 1093–1103.

121 *willing to pay the most:* The person who would have been willing to pay the most might fail to win the auction either (1) because he was outbid at the last second (although he would have raised his bid had there been more time, which is the whole point of sniping) or (2) because some sniped bid that would have won the auction was made too late and never recorded.

126 *Cook had no trouble:* Gareth Cook, "School Assignment Flaws Detailed," *Boston Globe,* September 12, 2003. See also Atila Abdulkadiroğlu and Tayfun Sönmez, "School Choice: A Mechanism Design Approach," *American Economic Review* 93, no. 3 (June 2003): 729–47.

127 *laboratory experiment:* Yan Chen and Tayfun Sönmez, "School Choice: An Experimental Study," *Journal of Economic Theory* 127 (2006): 202–31.

8. THE MATCH: STRONG MEDICINE FOR NEW DOCTORS

138 *might miss the chance:* Here's why. Consider a student who listed as his first choice a residency program to which he didn't match, but whose second-choice program included him among its first choices. That residency program might fill all its positions in the 1-1 and 2-1 steps of the algorithm and have no position available for our student, a 1-2 match. Thus it was possible for a student to actually be punished for ranking first a residency program to which he or she could not match by ending up at a program he liked very little—even though his second-choice program ranked him first.

141 *British clearinghouses:* On British hospitals, see A. E. Roth, "A Natural Experiment in the Organization of Entry-Level Labor Markets: Regional Markets for New Physicians and Surgeons in the U.K.," *American Economic Review* 81 (June 1991): 415–40.

a 1962 article: David Gale and Lloyd Shapley, "College Admissions and the Stability of Marriage," *American Mathematical Monthly* 69 (1962): 9–15.

143 *isn't at all obvious:* Just to see how easy it was to prove that result, let's prove the very same thing another way, starting with the residency program P. Suppose the medical staff members at P prefer some doctor (D) to one of the ones they were actually matched with. How do we know that D doesn't also prefer P? Because if P prefers D to someone it eventually hired, it must have made an offer to D first, since employers make offers in order of preference. And if P isn't matched to D, it's because he rejected P's offer when he got one that he preferred. He may have subsequently rejected that offer for one he liked even

better, but for sure the offer he finally accepted is one he liked better than P. So if P prefers D, we know that D doesn't prefer P. That is, whichever way you look at it, you can see that there aren't any docs and residency programs that aren't matched to each other but would prefer to be.

literal fanfares: You can hear the fanfare for Lloyd Shapley at the end of this two-minute video showing him receiving his Nobel Prize from the king of Sweden: http://www.nobelprize.org/mediaplayer/index.php?id=1906. (And here, moments earlier, you can hear mine: http://www.nobelprize.org/mediaplayer/index.php?id=1905.)

144 *finding the outcome . . . quickly:* The Match works quickly for two reasons: the participants decide on their preferences in advance, so that no one has to wait for someone else to decide, and the algorithm processes the "rejection chains" automatically, initially using card-sorting machines but today by computer. Both of these things are important. Xiaolin Xing and I studied the labor market for professional psychologists at a time when they tried to implement something like the deferred acceptance algorithm by telephone. It was too congested for them to get to a stable matching: trying to conduct all the steps of the deferred acceptance algorithm through long chains of phone calls took too long. See A. E. Roth and X. Xing, "Turnaround Time and Bottlenecks in Market Clearing: Decentralized Matching in the Market for Clinical Psychologists," *Journal of Political Economy* 105 (April 1997): 284–329. Today they use the same kind of computerized clearinghouse that we designed for the medical Match.

146 counterexample: Gale and Shapley's proof that a stable matching always exists when no couples are present is what mathematicians call a *theorem,* while an example that shows that conclusion no longer follows when couples are present is called a *counterexample,* because it is an example that goes counter to what we might have expected from the theorem. For this and other early observations about the medical match, see A. E. Roth, "The Evolution of the Labor Market for Medical Interns and Residents: A Case Study in Game Theory," *Journal of Political Economy* 92 (1984): 991–1016.

book on matching: Alvin E. Roth and Marilda A. Oliveira Sotomayor, *Two-Sided Matching: A Study in Game-Theoretic Modeling and Analysis* (Cambridge: Cambridge University Press, 1990).

147 *engineer:* For some further thoughts on economists as engineers, see Alvin E. Roth, "The Economist as Engineer: Game Theory, Experimentation, and Computation as Tools for Design Economics," *Econometrica* 70, no. 4 (July 2002): 1341–78, http://web.stanford.edu/~alroth/papers/engineer.pdf.

148 *partner in design:* When Muriel Niederle and I helped redesign the market for new gastroenterologists, Dr. Debbie Proctor at Yale was the inside champion and expert guide. In the chapters to come, I'll mention others.

Roth-Peranson algorithm: A. E. Roth and E. Peranson, "The Redesign of the Matching Market for American Physicians: Some Engineering Aspects of

Economic Design," *American Economic Review* 89, no. 4 (September 1999): 748–80.

149 *stable matchings will exist:* See Fuhito Kojima, Parag A. Pathak, and Alvin E. Roth, "Matching with Couples: Stability and Incentives in Large Markets," *Quarterly Journal of Economics* 128, no. 4 (2013): 1585–1632, and, for a subsequent stronger result, Itai Ashlagi, Mark Braverman, and Avinatan Hassidim, "Stability in Large Matching Markets with Complementarities," *Operations Research* 62, no. 4 (2014): 713–32.

Rural Hospitals Theorem: Alvin E. Roth, "On the Allocation of Residents to Rural Hospitals: A General Property of Two-Sided Matching Markets," *Econometrica* 54, no. 2 (1986): 425–27.

9. BACK TO SCHOOL

155 *a computerized clearinghouse:* Atila Abdulkadiroğlu, Parag A. Pathak, and Alvin E. Roth, "The New York City High School Match," *American Economic Review: Papers and Proceedings* 95, no. 2 (May 2005): 364–67.

158 *I made some simplifications:* For more details on New York City high school choice, see Atila Abdulkadiroğlu, Parag A. Pathak, and Alvin E. Roth, "Strategy-Proofness Versus Efficiency in Matching with Indifferences: Redesigning the NYC High School Match," *American Economic Review* 99, no. 5 (December 2009): 1954–78.

162 *Boston Public Schools:* For more details on Boston schools, see Atila Abdulkadiroğlu, Parag A. Pathak, Alvin E. Roth, and Tayfun Sönmez, "The Boston Public School Match," *American Economic Review: Papers and Proceedings* 95, no. 2 (May 2005): 368–71.

166 *clearinghouses have been modified:* For a description of Chinese college admissions, see Yan Chen and Onur Kesten, "From Boston to Chinese Parallel to Deferred Acceptance: Theory and Experiments on a Family of School Choice Mechanisms" (discussion paper, University of Michigan, 2013).

10. SIGNALING

170 *is the candidate* qualified: For signals about quality, see the Nobel Prize–winning paper by Michael Spence, "Job Market Signaling," *Quarterly Journal of Economics* 87, no. 3 (August 1973): 355–74.

171 *the Common App:* Incidentally, the Common App hasn't escaped having to deal with the congestion that so often comes with thickness. In 2013, its software suffered from overload, and the stress level among college applicants as deadlines approached made national news. This has sparked discussion among a coalition of colleges about developing a backup or alternative Internet application system.

admissions exams: Christopher Avery, Soohyung Lee, and Alvin E. Roth,

"College Admissions as Non-Price Competition: The Case of South Korea," NBER Working Paper No. 20774, December 2014.

American colleges: For more on college admissions, see Christopher Avery, Andrew Fairbanks, and Richard Zeckhauser, *The Early Admissions Game* (Cambridge, MA: Harvard University Press, 2003).

175 *"signaling mechanism":* On the economics job market, and on the mechanism we built to allow candidates to signal particular interest, see Peter Coles, John H. Cawley, Phillip B. Levine, Muriel Niederle, Alvin E. Roth, and John J. Siegfried, "The Job Market for New Economists: A Market Design Perspective," *Journal of Economic Perspectives* 24, no. 4 (Fall 2010): 187–206.

176 *The experiment allowed:* Soohyung Lee and Muriel Niederle, "Propose with a Rose? Signaling in Internet Dating Markets," *Experimental Economics,* forthcoming.

177 *And the effect of a rose:* This turns out to echo the effect of signals that we observe in the economics job market when we use the relative prestige of the university from which the applicant is graduating and the one to which he or she is applying as a measure of desirability.

178 *impressive genetic resources:* For signals of desirability in biology, see Amotz Zahavi, *The Handicap Principle: A Missing Piece of Darwin's Puzzle* (Oxford: Oxford University Press, 1997).

181 *one of the oldest:* Herodotus writes in *The Histories* (1.196) that the Babylonians used to sell marriageable girls, once a year, in an auction in which each of the most beautiful girls would be sold for a high price to the highest bidder among the wealthy men, and each of the others would go to the bidder who demanded the smallest dowry. This might be the oldest reference to a double auction, with both bids and asks, a little like those we see in the financial markets discussed in chapter 5.

184 *these two effects balance out:* The study of second-price auctions from this point of view won the 1996 Nobel Prize in Economics for William Vickrey, the author of the 1961 article "Counterspeculation, Auctions, and Competitive Sealed Tenders," *Journal of Finance* 16, no. 1 (March 1961): 8–37.

185 *auctions:* Auction design could fill a whole book, and in fact there are several on the subject. My Stanford colleague Paul Milgrom, the leader among modern auction designers, has written a book for economists titled *Putting Auction Theory to Work* (Cambridge: Cambridge University Press, 2004).

187 *"simultaneous ascending" auctions:* In 2014, Stanford professors Paul Milgrom and Bob Wilson and Microsoft chief economist Preston McAfee were awarded the Golden Goose Award for their work in designing the simultaneous ascending auction. The Golden Goose Award is given to highlight the human and economic benefits of federally funded research.

189 *lots of licenses for sale:* When there are *n* licenses for sale, there are 2*n*–1 distinct packages that could receive bids, so when there are 5 licenses, there are already 31 possible packages; when there are 10 licenses, there are 1,023

packages; and when there are 1,000 licenses, the number of packages takes hundreds of digits to write.

190 *conducted automatically by Google:* I've simplified a bit here. For one thing, Google gets paid when you click on the ad, so it doesn't look only at the advertiser's bid but also at how frequently the ad gets clicked on.

192 *even your emails:* In the early days of Gmail, a friend told me that he'd sent out an email to some friends saying, "Hey guys, anyone want to go out for Mexican food?" Soon after, he started seeing ads for "Mexican guys."

11. REPUGNANT, FORBIDDEN . . . AND DESIGNED

196 *repugnant transactions:* For more on repugnant transactions, see Alvin E. Roth, "Repugnance as a Constraint on Markets," *Journal of Economic Perspectives* 21, no. 3 (Summer 2007): 37–58, http://pubs.aeaweb.org/doi/pdfplus/10.1257/jep.21.3.37. For a variety of examples, see my market design blog: http://marketdesigner.blogspot.com/search/label/repugnance.

199 *same-sex marriage:* Here's a website that is keeping track of the status of same-sex marriage state by state: http://www.freedomtomarry.org/states/.

201 *"Now, how could":* Max Weber, *The Protestant Ethic and the Spirit of Capitalism,* trans. Talcott Parsons (Mineola, NY: Dover, 2003), 74.

203 *Another fear is* coercion: See Sandro Ambuehl, Muriel Niederle, and Alvin E. Roth, "More Money, More Problems? Can High Pay Be Coercive and Repugnant?," *American Economic Review, Papers and Proceedings* 105, no. 5 (May 2015), forthcoming.

206 *Iranian market:* For one description of the Iranian kidney market, see Sigrid Fry-Revere, *The Kidney Sellers: A Journey of Discovery in Iran* (Durham, NC: Carolina Academic Press, 2014).

allow living kidneys: See, for example, the passionate advocacy of Sally Satel, herself a kidney transplant recipient and a doctor, and the author of *When Altruism Isn't Enough: The Case for Compensating Kidney Donors* (Washington, DC: AEI Press, 2008), or the argument presented to economists by the late Nobel laureate Gary Becker and his coauthor Julio Elías in "Introducing Incentives in the Market for Live and Cadaveric Organ Donations," *Journal of Economic Perspectives* 21, no. 3 (Summer 2007): 3–24, http://pubs.aeaweb.org/doi/pdfplus/10.1257/jep.21.3.3.

"It is not": Adam Smith, *An Inquiry into the Nature and Causes of the Wealth of Nations* (Oxford: Oxford University Press, 2008), bk. 1, chap. 2, para. 2.

207 *opposition to legalizing kidney sales:* One of the most outspoken opponents of legalizing kidney sales is one of the heroes of kidney exchange, Frank Delmonico. He has been active in formulating the Declaration of Istanbul and is executive director of its custodian group, whose mission is "to promote, implement and uphold the Declaration of Istanbul so as to combat organ trafficking, transplant tourism and transplant commercialism and to encourage

adoption of effective and ethical transplantation practices around the world." See http://www.declarationofistanbul.org/.

211 *Nigeria:* For information about kidney disease in Africa, see Saraladevi Naicker, "End-Stage Renal Disease in Sub-Saharan Africa," *Ethnicity & Disease* 19, no. 1 (2009): 13.

12. FREE MARKETS AND MARKET DESIGN

219 *tablecloths:* I recall discussing tablecloths as an indicator of restaurant types years ago with my late colleague Gerald Salancik, at the University of Illinois.

226 The Road to Serfdom: All the quotes in this section are from F. A. Hayek, *The Collected Works of F. A. Hayek,* vol. 2, *The Road to Serfdom: Text and Documents — The Definitive Edition,* ed. Bruce Caldwell (Chicago: University of Chicago Press, 2007).

liberal: Hayek in fact wrote of the confusion of terms: "Indeed, what in Europe is or used to be called 'liberal' is in the USA today with some justification called 'conservative'; while in recent times the term 'liberal' has been used there to describe what in Europe would be called socialism. But of Europe it is equally true that none of the political parties which use the designation 'liberal' now adhere to the liberal principles of the nineteenth century." F. A. Hayek, "Liberalism," chap. 9 in *New Studies in Philosophy, Politics, Economics and the History of Ideas* (London: Routledge & Kegan Paul, 1982), 119–51.

227 *As in a garden:* When Eric Budish (about whom I wrote in chapter 5) speaks to audiences of finance professionals, some of whom have become billionaires, he often finds them initially skeptical about the idea that the markets they work in might profit from some judicious redesign. But he told me that after one talk, a famous financier stood up and said something like "I didn't expect to say this, but you're not a communist. Markets need rules, and you don't want more rules; you just want different rules."

229 *matching markets:* Of course, one of the contributions of market design is to bring into clear focus, as markets, a wider class of things than just commodity markets, in which price does all the work. The English language gives us a head start on that, speaking as it does not only of job markets but also of marriage markets.

Index